2nd International Combined Workshop on Spatial Language Understanding and Grounded Communication for Robotics (SpLU-RoboNLP 2021)

Held online due to COVID-19

Bangkok, Thailand
5 – 6 August 2021

ISBN: 978-1-7138-3386-4

Printed from e-media with permission by:

Curran Associates, Inc.
57 Morehouse Lane
Red Hook, NY 12571

CURRAN ASSOCIATES INC. proceedings.com

Some format issues inherent in the e-media version may also appear in this print version.

Association for Computational Linguistics
209 N. Eighth Street
Stroudsburg, Pennsylvania 18360

Phone: 1-570-476-8006
Fax: 1-570-476-0860

acl@aclweb.org

Additional copies of this publication are available from:

Curran Associates, Inc.
57 Morehouse Lane
Red Hook, NY 12571 USA
Phone: 845-758-0400
Fax: 845-758-2633
Email: curran@proceedings.com
Web: www.proceedings.com

SpLU-RoboNLP 2020

The 2nd International Combined Workshop on Spatial Language Understanding and Grounded Communication for Robotics

Proceedings of the Workshop

August 5-6, 2021
Bangkok, Thailand (online)

Introduction

SpLU-RoboNLP 2021 is a combined workshop on spatial language understanding (SpLU) and grounded communication for robotics (RoboNLP) that aims to realize the long-term goal of natural conversation with machines in our homes, workplaces, hospitals, and warehouses by highlighting developments in linking language to perception and actions in the physical world. It also highlights the importance of spatial semantics when it comes to communicating about the physical world and grounding language in perception. The combined workshop aims to bring together members of NLP, robotics, vision and related communities in order to initiate discussions across fields dealing with spatial language understanding and grounding language to perception and actions in the real world. The main goal of this joint workshop is to bring in the perspectives of researchers working on physical robot systems and with human users, and align spatial language understanding representation and learning approaches, datasets, and benchmarks with the goals and constraints encountered in HRI and robotics. Such constraints include high costs of real-robot experiments, human-in-the-loop training and evaluation settings, scarcity of embodied data, as well as non-verbal communication.

Recent years have seen an increase in the availability of simulators in which virtual agents can take actions and obtain realistic visual observations, which has led to the creating of benchmarks for grounded language understanding in such environments. These benchmarks allow more direct comparisons of different techniques on certain tasks and have led to a significant increase in interest in some tasks such as vision and language navigation. However, many challenges still remain. Most systems using such benchmarks do not actually perform interactive training - obtaining live feedback from the environment on taking novel actions. Such training becomes more expensive as the simulator starts to support more actions. Different simulators and benchmarks vary in the extent to which they model realistic tasks or realistic capabilities of physical robots. Many of the modeling techniques used on such benchmarks may require too much compute to be used on physical robots.

Following the exciting recent progress in a number of visual language grounding tasks and vision and language navigation, the creation of more interactive embodied agents that can reason about spatial knowledge, common sense knowledge and information provided in instructions, generalize to data beyond what is seen during training, identify gaps in their knowledge or understanding, and engage in natural language interactions with users to fill in these gaps and explain their behavior are interesting research directions.

We have accepted 6 archival submissions and the workshop included an additional 4 non archival submissions.

Organizers:

Malihe Alikhani, University of Pittsburgh
Valts Blukis, Cornell University
Parisa Kordjamshidi, Michigan State University
Aishwarya Padmakumar, Amazon Alexa AI
Hao Tan, University of North Carolina, Chapel Hill

Program Committee:

Jacob Arkin, University of Rochester
Jonathan Berant, Tel-Aviv University
Steven Bethard, University of Arizona
Johan Bos, University of Groningen
Volkan Cirik, Carnegie Mellon University
Guillem Collell, KU Leuven
Simon Dobnik, University of Gothenburg, Sweden
Fethiye Irmak Dogan, KTH Royal Institute of Technology
Frank Ferraro, University of Maryland, Baltimore County
Daniel Fried, University of California, Berkeley
Felix Gervits, Tufts University
Yicong Hong, Australian National University
Drew Arad Hudson, Stanford University
Xavier Hinaut, INRIA
Gabriel Ilharco, University of Washington
Siddharth Karamcheti, Stanford University
Hyounghun Kim, University of North Carolina, Chapel Hill
Jacob Krantz, Oregon State University
Stephanie Lukin, Army Research Laboratory
Lei Li, ByteDance AI Lab
Roshanak Mirzaee, Michigan State University
Ray Mooney, University of Texas, Austin
Mari Broman Olsen, Microsoft
Natalie Parde, University of Illinois, Chicago
Christopher Paxton, NVIDIA
Roma Patel, Brown University
Nisha Pillai, University of Maryland, Baltimore County
Preeti Ramaraj, University of Michigan
Kirk Roberts, University of Texas, Houston
Anna Rohrbach, University of California, Berkeley
Mohit Shridhar, University of Washington
Ayush Shrivastava, Georgia Tech
Jivko Sinapov, Tufts University
Kristin Stock, Massey University of New Zealand
Alane Suhr, Cornell University
Rosario Scalise, University of Washington
Morgan Ulinski, Columbia University
Xin Wang, University of California, Santa Cruz
Shiqi Zhang, SUNY Binghamton

Invited Speakers:

Maja Matarić, University of Southern California

Kartik Narasimhan, Princeton University
Jean Oh, Carnegie Mellon University
Thora Tenbrink, Bangor University

Table of Contents

Conference Program

Friday, Aug 6, 2021 (EDT)

09:00 - 10:00	**Poster Session**
10:00 - 12:00	**Morning Invited Talks**
10:00 - 11:00	*Invited Talk* Thora Tenbrink
11:00 - 12:00	*Invited Talk* Jean Oh
12:00 - 13:50	**Morning Session**
	Symbol Grounding and Task Learning from Imperfect Corrections Mattias Appelgren and Alex Lascarides
	Learning to Read Maps: Understanding Natural Language Instructions from Unseen Maps Miltiadis Marios Katsakioris, Ioannis Konstas, Pierre Yves Mignotte and Helen Hastie
	Visually Grounded Follow-up Questions: a Dataset of Spatial Questions Which Require Dialogue History Tianai Dong, Alberto Testoni, Luciana Benotti and Raffaella Bernardi
	Modeling Semantics and Pragmatics of Spatial Prepositions via Hierarchical Common-Sense Primitives Georgiy Platonov, Yifei Yang, Haoyu Wu, Jonathan Waxman, Marcus Hill and Lenhart Schubert
	Towards Navigation by Reasoning over Spatial Configurations Yue Zhang, Quan Guo and Parisa Kordjamshidi
	Learning to Parse Sentences with Cross-Situational Learning using Different Word Embeddings Towards Robot Grounding Subba Reddy Oota, Frederic Alexandre and Xavier Hinaut
	Error-Aware Interactive Semantic Parsing of OpenStreetMap Michael Staniek and Stefan Riezler
	Compositional Data and Task Augmentation for Instruction Following Soham Dan, Xinran Han and Dan Roth

14:00 - 15:00 **ACL Findings Papers**

Language-Mediated, Object-Centric Representation Learning
Ruocheng Wang, Jiayuan Mao, Samuel Gershman, Jiajun Wu

Probing Image-Language Transformers for Verb Understanding
Lisa Anne Hendricks, Aida Nematzadeh

Hierarchical Task Learning from Language Instructions with Unified Transformers and Self-Monitoring
Yichi Zhang, Joyce Chai

VLM: Task-agnostic Video-Language Model Pre-training for Video Understanding
Hu Xu, Gargi Ghosh, Po-Yao Huang, Prahal Arora, Masoumeh Aminzadeh, Christoph Feichtenhofer, Florian Metze, Luke Zettlemoyer

Grounding 'Grounding' in NLP
Khyathi Raghavi Chandu, Yonatan Bisk, Alan W Black

PROST: Physical Reasoning of Objects through Space and Time
Stéphane Aroca-Ouellette, Cory Paik, Alessandro Roncone, Katharina Kann

15:00 - 16:00 **Panel Session**

17:00 - 19:00 **Afternoon Invited Talks**

17:00 - 18:00 *Invited Talk*
Karthik Narasimhan

18:00 - 19:00 *Invited Talk*
Maja Mataric

19:00 - 20:15 **Afternoon Session**

Plan Explanations that Exploit a Cognitive Spatial Model
Raj Korpan and Susan L. Epstein

Fine-Grained Spatial Information Extraction in Radiology as Two-turn Question Answering
Surabhi Datta and Kirk Roberts

Interactive Reinforcement Learning for Table Balancing Robot
Haein Jeon, Yewon Kim and Bo-Yeong Kang

Multi-Level Gazetteer-Free Geocoding
Sayali Kulkarni, Shailee Jain, Mohammad Javad Hosseini, Jason Baldridge, Eugene Ie and Li Zhang

Interactive learning from activity description
Khanh Nguyen, Dipendra Misra, Robert Schapire, Miro Dudík and Patrick Shafto

20:15 - 21:00 **Poster Session**

Symbol Grounding and Task Learning from Imperfect Corrections

Mattias Appelgren
University of Edinburgh
mattias.appelgren@ed.ac.uk

Alex Lascarides
University of Edinburgh
alex@inf.ed.ac.uk

Abstract

This paper describes a method for learning from a teacher's potentially unreliable corrective feedback in an interactive task learning setting. The graphical model uses discourse coherence to jointly learn symbol grounding, domain concepts and valid plans. Our experiments show that the agent learns its domain-level task in spite of the teacher's mistakes.

1 Introduction

Interactive Task Learning (ITL) aims to develop agents that can learn arbitrary new tasks through a combination of their own actions in the environment and an ongoing interaction with a teacher (see Laird et al. (2017) for a survey). Because the agent continues to learn after deployment, ITL allows an agent to learn in an ever changing environment in a natural manner.

One goal of ITL is to have the interactions be as natural as possible for a human teacher, and many different modes of interaction have been studied: non-verbal through demonstration or teleoperation (Argall et al., 2009), or natural language: an embodied extended dialogue between teacher and agent, like between a teacher and apprentice. Our interest lies in natural language interactions where teachers can provide instructions (She et al., 2014), describe current states (Hristov et al., 2017) and define concepts (Scheutz et al., 2017), goals (Kirk and Laird, 2019), and actions (She et al., 2014), while the agent asks clarifying questions (She and Chai, 2017) and executes instructed commands. Teachers can also use corrective feedback (Appelgren and Lascarides, 2020). These approaches all assume that the teacher offers information that is both correct and timely. However, humans are error prone, and so in this paper we study how agents can learn successfully from corrective feedback even when the teacher makes mistakes.

Appelgren and Lascarides' model exploits *discourse coherence* (Hobbs, 1985; Kehler, 2002; Asher and Lascarides, 2003): that is, constraints on how a current move relates to its context. But their models assume that the teacher follows perfectly a specific dialogue strategy: she corrects a mistake as and when the agent makes it. However, humans may fail to perceive mistakes when they occur,. They also may, as a result, utter a correction much later than when the agent made the mistake, and thanks to the teacher being confident, but wrong, about the agent's capacity to ground NL descriptions to their referents, the agent may miscalculate which salient part of the context the teacher is correcting. In this paper, we present and evaluate an ITL model that copes with such errors.

In §2, we use prior work to motivate the task we tackle, as described in §3. We present our ITL model in §4 and §5, focusing on coping with situations where the teacher makes mistakes of the type we just described. We show in §6 that by making the model separate the appearance that the teacher's utterance coherently connects to the agent's latest action with the chance that it is *not* so connected, our agent can still learn its domain-level task effectively.

2 Background

Interactive Task learning (ITL) exploits interaction to support autonomous decision making during planning (Laird et al., 2017). Similar to Kirk and Laird (2019), our aim is to provide the agent with information about goals, actions, and concepts that allow it to construct a formal representation of the decision problem, which can thereafter be solved with standard decision making algorithms. Rather than teaching a specific sequence of actions (as in e.g., Nicolescu and Mataric (2001); She et al. (2014)), the teacher provides the infor-

Proceedings of Second International Combined Workshop on Spatial Language Understanding andGrounded Communication for Robotics, pages 1–10
August 5–6, 2020. ©2021 Association for Computational Linguistics

mation needed to infer a valid plan for a goal in a range of specific situations. In this work we focus on learning goals, which express constraints on a final state. The agent learns these goals by receiving *corrective* dialogue moves that highlight an aspect of the goal which the agent has violated (Appelgren and Lascarides, 2020).

Natural language (NL) can make ITL more data efficient than non-verbal demonstration alone: even simple yes/no feedback can be used to learn a reward function (Knox and Stone, 2009) or to trigger specific algorithms for improving behaviour (Nicolescu and Mataric, 2003). More extended NL phrases must map to semantic representations or logical forms that support inference (eg, Wang et al., 2016; Zettlemoyer and Collins, 2007). Like prior ITL systems, (eg, Forbes et al., 2015; Lauria et al., 2002) we assume our agent can analyse sentential syntax, restricting the possible logical forms to a finite set. But disambiguated syntax does not resolve semantic scope ambiguities or lexical senses (Copestake et al., 1999), and so the agent must use context to identify which logical form matches the speaker's intended meaning.

Recovering from misunderstandings has been addressed in dialogue systems (eg, Skantze, 2007), and ITL systems cope with incorrect estimates of denotations of NL descriptions (eg, Part and Lemon, 2019). Here, we address new sources of misunderstanding that stem quite naturally from the teacher attempting, but failing, to abide by a particular dialogue strategy: ie, to correct the agent's mistakes as and when they're made. This can lead to the learner misinterpreting the teacher's silence (silence might *not* mean the latest action was correct) or misinterpreting which action is being corrected (it might be an earlier action than the agent's latest one). We propose a model that copes with this uncertainty.

3 Task

In our task an agent must build towers in a blocks world. The agent begins knowing two PDDL action descriptions: $put(x, y)$ for putting an object x on another y; and $unstack(x, y)$ for removing an object x from another object y and placing x back on the table. Further, it knows the initial state consists of 10 individual blocks that are clear (i.e., nothing on them) and on the table, and that the goal G contains the fact that all the 10 blocks must be in a tower.

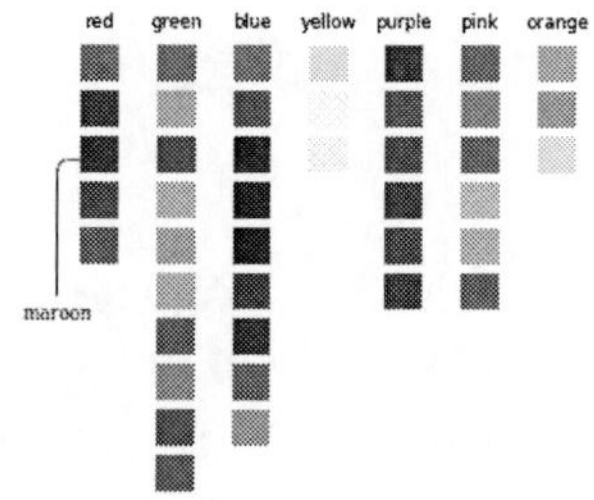

Figure 1: The colours of objects fit into different colour terms. Each individual hue is generated from a Gaussian distribution, with mean and variance selected to produce hues described by the chosen colour term. There are high level categories like "red" and "green" and more specific ones like "maroon". This figure shows examples of hues generated in each category, including one that is both red and maroon.

However, putting the blocks in a tower is only a partial description of the true planning problem, and the agent lacks vital knowledge about the problem in the following ways. First, the true goal G includes further constraints (e.g., that each red block must be on a blue block) and the agent is unaware of which such constraints are truly in the goal. Further, and perhaps more fundamentally, the agent is also unaware of the colour terms used to define the constraints. I.e. the word "red" is not a part of the agent's natural language vocabulary, and so the agent does not know what "red" means or what particular set of RGB values the word denotes. Instead, the agent can only observe the RGB value of an object and must learn to recognise the colour through interaction with the teacher, and in particular the corrective dialogue moves that the teacher utters.

The possible goal constraints are represented in equations (1–2), where C_1 and C_2 are colour terms; e.g., "red" (r for short) and "blue" (b).

$$r_1^{c_1,c_2} = \forall x.c_1(x) \rightarrow \exists y.c_2(y) \wedge on(x,y) \quad (1)$$

$$r_2^{c_1,c_2} = \forall y.c_2(y) \rightarrow \exists x.c_1(x) \wedge on(x,y) \quad (2)$$

In words, $r_1^{r,b}$ expresses that every red block must be on a blue block; $r_2^{r,b}$ that every blue block should have a red one on it. These rules constrain the final tower, but thanks to the available actions, if a constraint is violated by a put action then it remains violated in all subsequent states unless that put action is undone by $unstack$.

In our experiments (see §6), a simulated teacher observes the agent attempting to build the tower, and when the agent executes an action that breaks one or more of the rules in the goal G, the teacher provides NL feedback—e.g., "no, red blocks should be on blue blocks". The feedback corrects the agent's action and provides an explanation as to why it was incorrect. However, linguistic syntax makes the sentence *ambiguous* between two rules—"red blocks should be on blue blocks" could mean $r_1^{r,b}$ or $r_2^{r,b}$. Thus, the agent must disambiguate the teacher's message while simultaneously learning to ground new terms in the embodied environment, in this example the terms "red" and "blue". This latter task amounts to learning which RGB values are members of which colour concepts (see Figure 1).

4 Coherence

To learn from the teacher's feedback the agent reasons about how an utterance is coherent. In discourse each utterance must connect to a previous part of the discourse through a coherence relation, and the discourse relation which connects the two informs us what the contribution adds to the discourse. In our multimodal discourse each of the teacher's utterances u connect to one of the agent's actions a through the discourse relation "correction". The semantics of correction stipulate that the content of the correction is true and negates some part of the corrected action (Asher and Lascarides, 2003). In our domain, this means that the teacher will utter u if the agent's latest action a violates the rule that she intended u to express. If $u =$ "no, red blocks should be on blue blocks" then, as previously stated, this is ambiguous between is $r_1^{r,b}$ and $r_2^{r,b}$. So, a must violate one of these two rules:

$$CC(a,u) \leftrightarrow (r_1^{r,b} \in G \wedge V(r_1^{r,b},a)) \vee \\ (r_2^{r,b} \in G \wedge V(r_2^{r,b},a)) \quad (3)$$

where $CC(a,u)$ represents that u coherently corrects action a, G is the (true) goal, and $V(r_1^{r,b},a)$ represents that a violated $r_1^{r,b}$ (similarly for $V(r_2^{r,b},a)$). Since the semantics of correction is satisfied only if the correction is true, the rule the speaker intended to express must also be part of the true goal G; that is why (3) features $r_1^{r,b} \in G$ (and $r_2^{r,b} \in G$) in the two disjuncts.

There are two ways in which these rules can be violated. Either directly or indirectly. For $r_1^{r,b}$ the rule requires every red block to be on a blue block, therefore it is directly violated by action $a = put(o_1, o_2)$ if o_1 is red and o_2 is not blue (illustrated in S_1 of Figure 2):

$$V_D(r_1^{r,b},a) \leftrightarrow red(o_1) \wedge \neg blue(o_2) \quad (4)$$

The rule $r_2^{r,b}$ requires all blue blocks to have red blocks on them, meaning that S_1 in Figure 2 does not directly violate the rule, but S_2 does because it is only violated when a blue block does not have a red block on it:

$$V_D(r_2^{r,b},a) \leftrightarrow \neg red(o_1) \wedge blue(o_2) \quad (5)$$

So $r_1^{r,b}$ is not directly violated in S_2 and $r_2^{r,b}$ is not directly violated in S_1 but it would still be impossible to complete a rule compliant tower without undoing the progress that has been made on tower building. This is because the block which is currently not in the tower cannot be placed into the current tower in a rule compliant manner. For $r_1^{r,b}$ in S_2 the red block needs a blue block to be placed on, but no such blue block exists. Similarly, for $r_2^{r,b}$ in S_1 the blue block needs a red block to place on it, but no additional red blocks are available. In this way the rules are Indirectly violated in these states, which occurs when the number of available blocks of each colour makes it impossible to place all of those blocks:

$$V_I(r_1^{r,b},a) \leftrightarrow \neg red(o_1) \wedge blue(o_2) \\ |\{o_3 : red(o_3) \wedge on(o_3, table)\}| > \\ |\{o_4 : blue(o_4) \wedge on(o_4, table)\}| \quad (6)$$

$$V_I(r_2^{r,b},a) \leftrightarrow red(o_1) \wedge \neg blue(o_2) \wedge \\ |\{o_3 : blue(o_3) \wedge on(o_3, table)\}| > \\ |\{o_4 : blue(o_4) \wedge on(o_4, table)\}| \quad (7)$$

Our teacher signals if the error is due to a direct violation V_D by pointing at the tower or an indirect violation V_I by pointing at a block which cannot be placed in the tower any more (e.g., the blue block in S_1 or the red block in S_2).

When the agent observes the teacher say $u =$ "no, put red blocks on blue blocks" it can make inferences about the world, with confidence in those inferences depending on its current knowledge. For example, if it knows with confidence which blocks are "red" or "blue", then it can infer via equations (4–7) which of the rules the teacher intended to convey. Alternatively, if the agent knows which rule was violated then the agent can infer the colour of the blocks. We use this in §5 to learn the task. However, if the agent is completely ignorant about

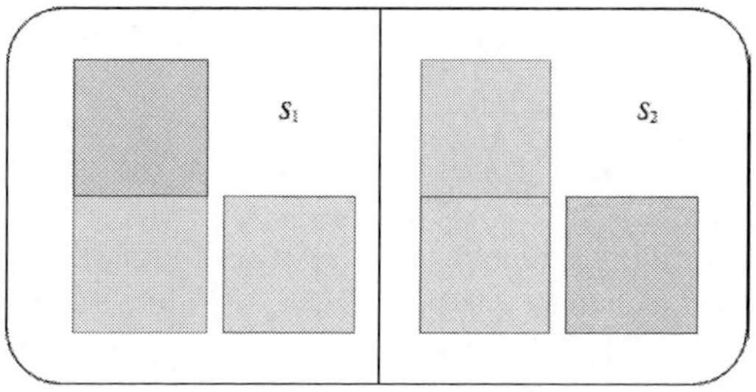

Figure 2: These two states would both be corrected if either $r_1^{(r,b)}$ or $r_2^{(r,b)}$ were in the goal.

the referents of the colour terms, it may not be able to make an inference at all. In this case it will ask for help by asking the colour of one of the blocks: “is the top block red?”. Either answer to this question is suffiicent for the agent to disambiguate the intended message and also gain training exemplars (both positive and negative) for grounding the relevant colour terms.

If the teacher’s dialogue strategy is to always correct an action which violates a rule (either directly or indirectly) directly after this incorrect action is executed, then the teacher’s silence implies that the latest executed action does not violate a rule. This means that if the agent knows, for example, that if a green block is placed on a blue block then either green blocks must always be placed on blue blocks ($r_1^{g,b}$) or no rule constraining green blocks exists. In this way the teacher’s silence implies assent.

4.1 Faulty Teacher

We’ve laid out the what it means for something to be coherent, assuming that the teacher always acts in the most optimal way, correcting any action which violates a rule as soon as that action is performed. However, in general a human teacher will be unlikely to perfectly follow this strategy. Despite this, an agent would still have to attempt to learn from the teacher’s utterances even though some of those utterances may not fit with the agent’s expectations and understanding of coherence. In this paper we introduce two types of errors the teacher can make: (a) she fails to utter a correction when the latest action a violates a rule; and (b) she utters a correction when the most recent action does not violate a rule (perhaps because she notices a previous action she should have corrected). We think of (b) as *adding* a correction at the ‘wrong’ time.

Since a rule can violate an action in two ways—either Directly or Indirectly—teacher errors of type (a) and (b) lead to four kinds of ‘imperfect’ dialogue moves:

1. Missing Direct Violations (MD)
2. Adding Direct Violations (AD)
3. Missing Indirect Violations (MI)
4. Adding Indirect Violations (AI)

In our experiments we control in what way the teacher is faulty by assigning a probability with which the teacher performs each type of mistake, e.g. P_{MD} represents the probability that the teacher misses a direct violation. Controlling these probabilities allows us to create different types of faulty teachers.

Due to the teacher’s faultiness the agent must now reason about whether or not it should update its knowledge of the world given a teacher utterance or silence. In the following section we describe how we deal with this by creating graphical models which capture the semantics of coherence as laid out in this section.

5 System Description

An agent for learning from correction to perform the task described in §3 must be able to update its knowledge given the corrective feedback and then use that updated knowledge to select and execute a plan. The system we have built consists of two main parts: Action Selection and Correction Handling.

5.1 Action Selection

To generate a valid plan, the agent uses the MetricFF symbolic planner (Hoffmann and Nebel, 2001; Hoffmann, 2003)). It requires as input a representation of the current state, the goal, and the action descriptions (here, `put(x, y)` and `unstack(x, y)`. The agent knows the position of objects, including which blocks are on each other, and it knows that the goal is to build a tower. However, the agent begins unaware of predicate symbols such as `red` and `blue` and ignorant of the rules $r \in G$ that constrain the completed tower.

The aim of our system is to learn to recognise the colours—and so estimate the current state S^*—and to identify the correct goal G, given the evidence $\mathbf{X}$ which it has observed so far. We describe how shortly. The agent uses its current knowledge to construct S^* and G which are given as input to the planner to find a valid plan. Due to errors in the grounding models or goal estimate, this may fail: eg, if the agent estimates $r_1^{r,b} \in G$ but there are more red blocks than blue blocks in S^*, making it impossible to place all of the red blocks. In such cases, the agent recovers by searching in the probabilistic neighbourhood of S^* for alternatives from which a valid plan for achieving G can be constructed (Appelgren and Lascarides, 2020). The agent executes each action in its plan until it's completed or the teacher gives corrective feedback. The latter triggers the Correction Handling system (see §5.2).

5.1.1 Grounding Models

The grounding models construct a representation of the current state S^* by predicting the colour of blocks, given their visual features. Binary classifiers represent the probability of an object being a particular colour, e.g. $P(Red_x|F_x)$ where F_x are the visual features of object x. We use binary classifiers over a categorical distribution for *colour* since the set of possible colours is unknown and colours aren't all mutually exclusive (e.g., maroon and red). We estimate the the probability using Bayes Rule:

$$P(Red_x)|F_x) = \frac{P(F_x|Red_x)P(Red_x)}{\sum_{i\in\{0,1\}} P(F_x|Red_x = i)P(Red_x = i)} \quad (8)$$

For $P(F_x|Red_x = 0)$ we use a uniform distribution—we expect colours that are not red to be distributed over the entire spectrum. $P(F_x|Red_x = 1)$ is estimated with weighted Kernel Density Estimation (wKDE). wKDE is a non-parametric model that puts a kernel around every known data point $\{(w_1, F_{x1}), ...(w_m, F_{xm})\}$ (where w_i are weights) and calculates the probability of a new data point via a normalised weighted sum of the values of the kernels at that point. With kernel φ (we use a diagonal Gaussian kernel), this becomes:

$$P(F_x|Red_x = 1) = \frac{1}{\sum_{i=1}^{m} w_i} \sum_{i=1}^{m} w_i \cdot \varphi(F_x - F_{x_i}) \quad (9)$$

The pairs (w_i, F_{x_i}) are generated by the Correction Handling system (see §5.2).

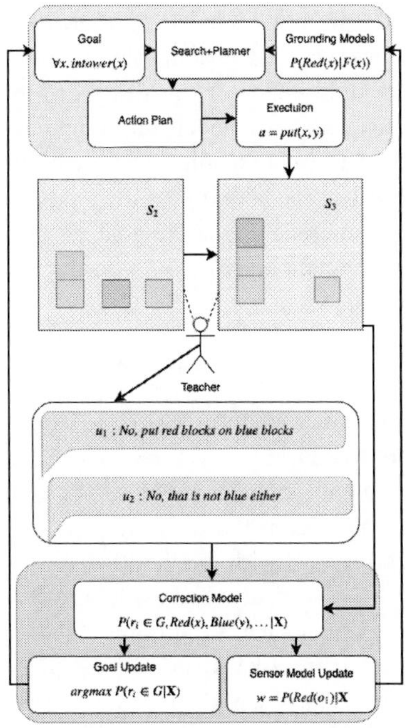

Figure 3: The agent consists of an Action Selection system (yellow) and a Learning System (green). The former uses a symbolic planner to find a plan given the most likely goal and symbol grounding. The latter uses coherence to learn the goal and grounding.

5.1.2 The Goal

In order to estimate G the agent begins with the (correct) knowledge that it must place all blocks in a tower. However, it must use the teacher's feedback $\mathbf{X}$ to find the most likely set of additional rules which are also conjuncts in G (see §5.2):

$$G = \arg \max_{r_1,\ldots,r_n} P(r_1 \in G, \ldots, r_n \in G|\mathbf{X}) \quad (10)$$

$\mathcal{R} = \{r_1 \ldots r_n\}$ is the set of possible rules that the agent is currently aware of, as determined by the colour terms it's aware of (so $\mathcal{R}$ gets larger during learning). For each $r \in \mathcal{R}$, the agent tracks its probabilistic belief that $r \in G$. Due to the belief that any one rule being in the goal is unlikely, the priors for all $r \in G$ are low: $P(r \in G) = 0.1$. And due to the independence assumption (11), the goal G is constructed by adding $r \in \mathcal{R}$ as a conjunct iff $P(r \in G|\mathbf{X}) > 0.5$.

$$P(r \in G, r' \in G|\mathbf{X}) = P(r \in G|\mathbf{X})P(r' \in G|\mathbf{X}) \quad (11)$$

5.2 Handling Corrections

When the teacher corrects the agent by uttering, for example, u = "no, red blocks should be on blue blocks" the agent must update its knowledge of the world in two ways: it must update its beliefs about what rules are in the goal, as described in §5.1.2 and it must update its models for grounding colour terms. To perform these inferences the agent builds a probabilistic model which allows it to perform these two inference. For the goal it uses the inference in equation (10). To learn the colours it performs this inference:

$$w = P(Red(o_1)|\mathbf{X}) \tag{12}$$

And adds the data point $(w, F(o_1))$ to its grounding model for red objects.

We base our graphical model on the model presented in Appelgren and Lascarides (2020) which we extend to deal with the fact that the teacher's utterance may be faulty. The model is a Bayes Net consisting of a number off different factors which are multiplied together to produce the final output probability. The model from Appelgren and Lascarides (2020) is shown in Figure 4. Grey nodes are observable while white nodes are latent. Arrows show conditional dependence between nodes. If the teacher is faultless then the agent observes that a coherent correction occurred: $CC(a, u)$. The factor for this in the graphical model:

$$P(CC(a,u)|r_1^{r,b} \in G, V(r_1^{r,b}, a), r_2^{r,b} \in G, V(r_2^{r,b}, a)) \tag{13}$$

captures equation (3), which stipulates that a coherent correction occurs when a rule which is in the goal is violated. In the graphical model the factor has a value of 1 any time this is true and 0 otherwise.

The violation factors $V(r_1^{r,b}, a)$ and $V(r_2^{r,b}, a)$ represent whether or not a particular rule was violated by the action a. The agent cannot observe this directly, but must instead infer this from whether or not the objects are red and blue. As such the factor:

$$P(V(r_i^{r,b}, a)|Red_{o_1}, Blue_{o_2}) \tag{14}$$

captures equation (4) for $i = 1$ and (5) for $i = 2$. The value of the factor is 1 if the relevant equation holds and 0 otherwise. So, for example, when $V(r_1^{r,b}, a) = True$, $Red_{o_1} = True$, and $Blue_{o_1} = False$ the value of the factor is 1.

The remaining nodes $P(Red_{o_1}|F_{o_1})$ and $P(Blue_{o_2}|F_{o_2})$ are defined by the agent's grounding models. $P(F_{o_i})$ is a prior which is assumed to be a constant for all o_i. Finally, $P(r_i^{r,b} \in G)$ is the agent's prior belief that $r_i^{r,b}$ is in the goal ($i = 1, 2$). As we mentioned earlier, this is initially set to 0.1; however, the prior is updated each time the agent encounters a new planning problem instance. The prior is then set simply to the agent's current belief given the available evidence.

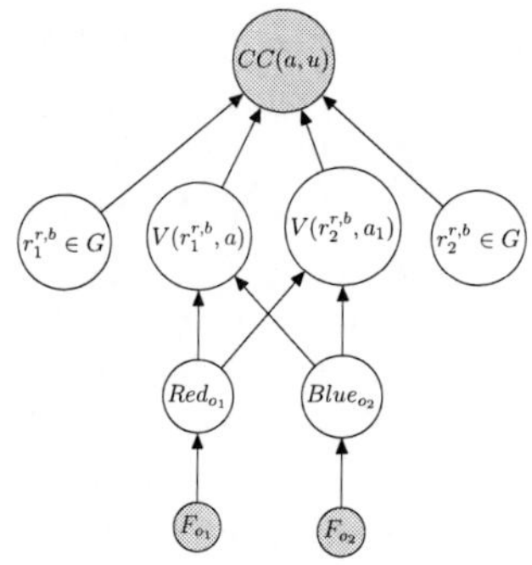

Figure 4: The nodes added to the graphical model after a correction u = "no, red blocks should be on blue blocks".

When the teacher designates a block o_3 on the table (thereby signaling that violation is indirect), the graphical model this generates is similar to Figure 4, save there are two additional nodes F_{o_3} and $Red_{o_3} \vee Blue_{o_3}$ (see (Appelgren and Lascarides, 2020) for details).

When the teacher stays silent the agent can make an inference which implies that no rule which is in the goal was violated. It can therefore build a model similar to Figure 4 which captures this negation of equation (3). The agent can then update its knowledge by making the same inferences when a correction occurs, but with the observed evidence being that no correction occurred. For further details on how this inference works see (Appelgren and Lascarides, 2020).

5.3 Uncertain Inferences

In this paper we assume that the teacher may make mistakes as described in §4. This introduces a novel problem for the agent since it can no longer assume that when the teacher says u that that means the utterance coherently attaches to the most recent action a. In other words, $CC(a, u)$ becomes latent, rather than observable. What *is* observable is that the teacher did in fact utter correction u immediately after action a. We capture this by adding a new (observable) factor $TeacherCorrection(a, u)$ (or $TC(a, u)$ for short) to the graphical model. When the teacher is infallible $TC(a, u) \equiv CC(a, u)$ but not when the teacher is fallible.

The updated model is shown in Figure 5. $TC(a, u)$ is added as an observable node with $CC(a, u)$ made latent. The factor for $CC(a, u)$ still works in the same way as before, conforming to equation (3). $TC(a, u)$ imposes no semantic constraints on a or on u. However, we can use the evidence of $TC(a, u)$ to inform the agent's belief about whether $CC(a, u)$ is true or not, i.e. whether it was actually coherent to utter u in response to a. The newly added factor $P(TC(a, u)|CC(a, u)$ captures the agent's belief about how faulty the teacher is and allows the agent to therefore reason about whether $TC(a, u)$ actually means that $CC(a, u)$. In essence, it answers the question "if it is coherent to correct a with u, how likely is it that the teacher actually says u". So, if the agent believes that the teacher forgets to utter a correction with probability $p = 0.1$ then $P(TC(a, u) = False|CC(a, u) = True) = 0.1$. Or if the agent believes that the teacher will falsely correct an action which was actually correct 5% of the time then $P(TC(a, u) = True|CC(a, u) = False) = 0.05$. This allows the agent to make use of the fact that the teacher did (or didn't) utter something to still update its beliefs about which rules are in the goal and what the colour of objects are.

The agents beliefs about the teacher's fallibility could be estimated from data or could potentially be updated on the fly given the agent's observation of the interaction. However, for the purpose of the experiments in this paper we have direct access to the true probability of teacher fallibility since we explicitly set this probability ourselves. We therefore set the agent's belief about the teacher's fallibility to the true value.

The final change made to the system compared to Appelgren and Lascarides (2020) is to the way inference is done. In their paper they perform exact inference in a manner which was optimised for the structure of the graphical model and the incremental nature of the inference. However, the method relied on the fact that the majority of probability states had zero probability due to the deterministic factors in the model. When the teacher is fallible the number of zero probability states greatly falls. This leads to a situation where exact inference becomes impractical. To deal with this we deploy approximate inference, based on

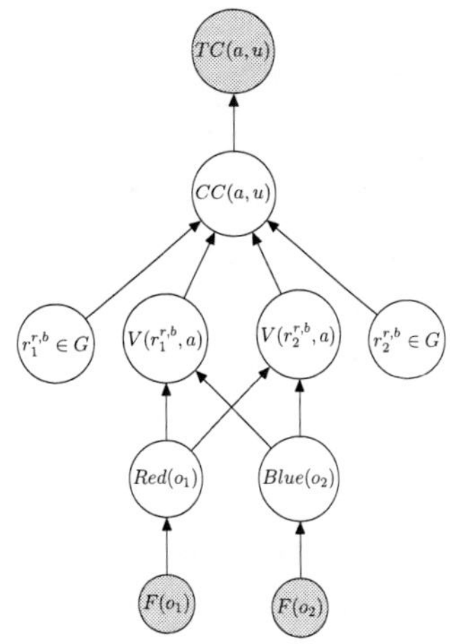

Figure 5: The nodes added to the graphical model after a correction u = "no, red blocks should be on blue blocks". Grey is observed and white latent.

a simple Bayesian Update together with a beam search method which relies on the fact that the model grows incrementally. We first find the probability for every atomic state in the newly added model chunk. This establishes a set of possible non-zero probability atomic states. These are combined with atomic states from the previous inference steps which we call the beam. The beam is the N most likely states from the previous state. Each new non-zero atomic state is combined with states from the beam if they are compatible, determined by both states having the same value for any overlapping variables. These new combined atomic states are evaluated on the full model and the N most likely are kept as a new beam, which is normalised to create a consistent probability distribution. Specific probabilities can then be calculated by summing all atomic states that match the chosen value, eg, where $Red_{o_1} = True$.

6 Experiments

In §4 we mentioned four types of teacher error and in our experiments we vary the level of the teacher's error in these different types. We believe the most likely is missing indirect errors (MI) since spotting these requires search on all possible future actions. So our first faulty teacher varies $P_{MI} \in \{0.0, 0.1, 0.25, 0.5, 1.0\}$: ie, from no errors to never correcting any indirect violations at all. Our second teacher makes mistakes with direct violations. We believe missing and adding direct violations will be linked, so we experiment

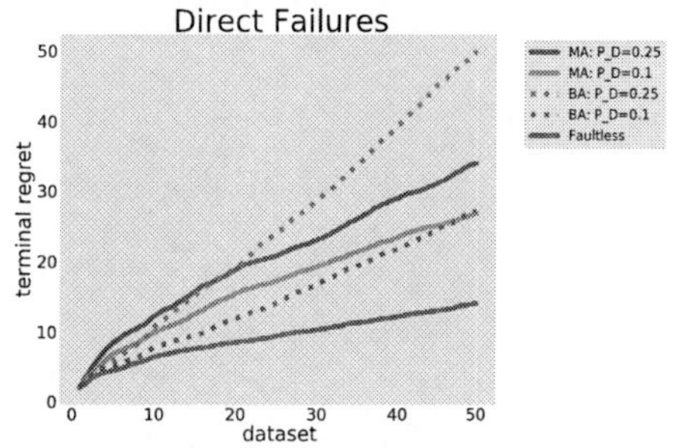

Figure 6: Cumulative regret for teachers making mistakes with direct violations. The dotted lines show the baseline agent while the solid lines show the mistake aware agent.

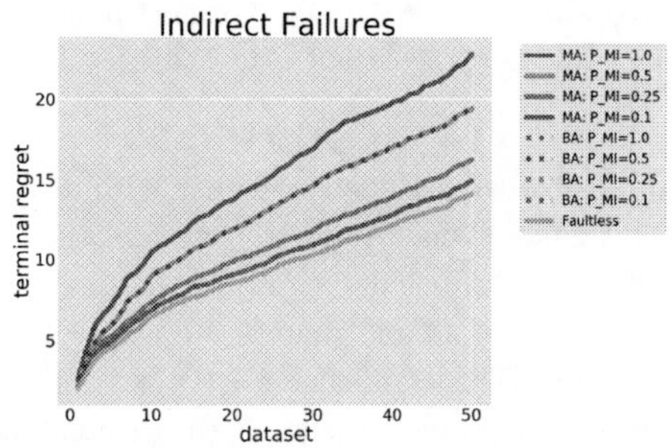

Figure 7: Cumulative regret for teachers making mistakes with indirect violations. The dotted lines show the baseline agent while the solid lines show the mistake aware agent.

with: $P_D = P_{MD} = P_{AD} \in \{0.0, 0.1, 0.25\}$. We study two types of agents in our experiments. The first, baseline agent, *BA*, ignores the fact that the teacher may be faulty. It simply uses the model described in Appelgren and Lascarides (2020). The only difference is that since the teacher is actually making mistakes, sometimes the agent may be given contradictory evidence which would cause the inference to fail. In such a situation the agent would simply ignore everything that was said in the current scenario and move on to the next planning problem instance. The second agent is a mistake-aware agent, *MA*, which makes inferences using the model from §5.3, matching its belief about the teacher's faultiness to the true probability.

In our experiments each agent is given 50 planning problems. Each planning problem has a different goal and a different set of 50 planning problem instances. The agent is reset between each planning problem, but retains knowledge between the 50 problem instances. We measure the number of mistakes the agent makes, which we call regret. A mistake is counted when an action takes a tower from a state where it is possible to complete it in a rule compliant way to one where it isn't without unstacking blocks. In Figures 6 and 7 we present the mean performance over the 50 planning problems, and we use paired t-tests to establish significance.

Let's begin by looking at the results for agents learning from teachers that fail to make corrections for indirect violations, shown in Figure 7. Clearly when the teacher is faulty the agent performs worse (a result which is shown significant through a pairwise t-test and significance threshold $p < 0.01$). However, two interesting things can be observed. First, the slope of the curves are about the same for the agents learning from the faulty teacher and those learning from the faultless teacher. What this implies is that although the agent takes longer to learn the task when the teacher misses indirect violations it does seem to reach an equal proficiency by the end. We can explain the fact that the agent makes more mistakes by the fact that it is unaware of several mistakes it is making, however, when it is made aware of a mistake it still manages to learn. The second point is that the BA and MA agents are equally good at learning at all levels of teacher error. There is a good reason for this. When the teacher misses indirect violations the agent can actually trust all other information it receives. If it is given a direct correction then it knows for certain that the teacher give a coherent correction. This is true for all the feedback the agent receives when the only error the teacher makes is missing indirect violations. For this reason there isn't actually any need to change the way in which the agent learns, which is reassuring given that we believe the indirect violations to be more likely to happen in practice.

Looking at the results when the teacher will both miss and add corrections for direct violations, shown in Figure 6, we see that the agent's performance is much worse, both compared to the faultless agent and to the agents learning from the teachers making direct violations (these results are also significant given a pairwise t-test and significance threshold $p < 0.01$). The big difference in this case is that the agent *BA* performs much worse than *MA*, especially when the likelihood of failure is higher. This is true both if we look at the final number of mistakes, but also at the slope of the

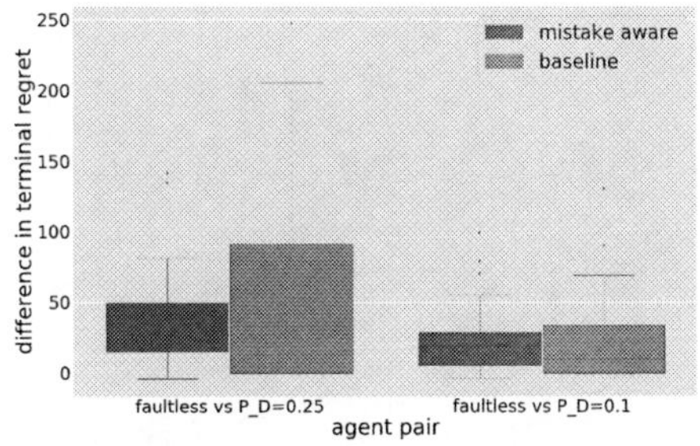

Figure 8: The difference in terminal regret when dealing with a faulty teacher vs. a faultless teacher, comparing the baseline *BA* to the mistake-aware agent *MA*.

curve, indicating that the agent is still making more mistakes by the end of training. Figure 8 shows why: it compares the difference between the terminal regret for the faultless teacher vs. the faulty one. For *BA* there is a much larger spread of outcomes, with a long tail of very high regrets. The results for *MA* reside in a much narrower region. This implies that in contrast to *MA*, *BA* performs extremely badly in a significant number of cases. The high regret scenarios can be explained by situations where the agent has failed to learn the task successfully and is therefore acting almost randomly. So, making the agent mistake-aware stabilises the learning process, allowing the agent to recover from the teacher's mistakes without completely failing to learn the task, as seen in the baseline.

7 Conclusion

In this paper we present an ITL model where the agent learns constraints and concepts in a tower building task from a teacher uttering corrections to its actions. The teacher can make mistakes, and to handle this we introduce a separation between the teacher uttering a correction (observable) vs. whether that correction coherently relates to the latest action (latent). Our experiments showed that this separation significantly reduces the proportion of situations where the agent fails to learn; without the separation, learning can go catastrophically wrong when the teacher's mistakes involve direct violations.

Acknowledgements: We thank the UKRI-funded TAS Governance Node (grant number EP/V026607/1) for their support, and three anonymous reviewers for helpful feedback. All remaining errors are our own.

References

Mattias Appelgren and A. Lascarides. 2020. Interactive task learning via embodied corrective feedback. *Auton. Agents Multi Agent Syst.*, 34:54.

Brenna Argall, Sonia Chernova, Manuela Veloso, and Brett Browning. 2009. A survey of robot learning from demonstration. *Robotics and Autonomous Systems*, 57:469–483.

Nicholas Asher and Alex Lascarides. 2003. *Logics of conversation*. Cambridge University Press.

Ann Copestake, Dan Flickinger, and Ivan A. Sag. 1999. Minimal recursion semantics: An introduction.

Maxwell Forbes, Rajesh P. N. Rao, Luke Zettlemoyer, and Maya Cakmak. 2015. Robot programming by demonstration with situated spatial language understanding. In *IEEE International Conference on Robotics and Automation, ICRA 2015, Seattle, WA, USA, 26-30 May, 2015*, pages 2014–2020.

J. R. Hobbs. 1985. On the coherence and structure of discourse. Technical Report csli-85-37, Center for the Study of Language and Information, Stanford University.

Jörg Hoffmann. 2003. The Metric-FF planning system: Translating "ignoring delete lists" to numeric state variables. 20:291–341.

Jörg Hoffmann and Bernhard Nebel. 2001. The FF planning system: Fast plan generation through heuristic search. 14:253–302.

Yordan Hristov, Svetlin Penkov, Alex Lascarides, and Subramanian Ramamoorthy. 2017. Grounding symbols in multi-modal instructions. In *Proceedings of the First Workshop on Language Grounding for Robotics, RoboNLP@ACL 2017, Vancouver, Canada, August 3, 2017*, pages 49–57.

A. Kehler. 2002. *Coherence, Reference and the Theory of Grammar*. csli Publications, Cambridge University Press.

James R. Kirk and John E. Laird. 2019. Learning hierarchical symbolic representations to support interactive task learning and knowledge transfer. In *Proceedings of the Twenty-Eighth International Joint Conference on Artificial Intelligence, IJCAI 2019, Macao, China, August 10-16, 2019*, pages 6095–6102.

W. Bradley Knox and Peter Stone. 2009. Interactively shaping agents via human reinforcement: the TAMER framework. In *Proceedings of the 5th International Conference on Knowledge Capture (K-CAP 2009), September 1-4, 2009, Redondo Beach, California, USA*, pages 9–16.

John E. Laird, Kevin A. Gluck, John R. Anderson, Kenneth D. Forbus, Odest Chadwicke Jenkins, Christian Lebiere, Dario D. Salvucci, Matthias Scheutz, Andrea Lockerd Thomaz, J. Gregory Trafton, Robert E.

Wray, Shiwali Mohan, and James R. Kirk. 2017. Interactive task learning. *IEEE Intelligent Systems*, 32:6–21.

Stanislao Lauria, Guido Bugmann, Theocharis Kyriacou, and Ewan Klein. 2002. Mobile robot programming using natural language. *Robotics and Autonomous Systems*, 38(3-4):171–181.

Monica N. Nicolescu and Maja J. Mataric. 2001. Experience-based representation construction: learning from human and robot teachers. In *IEEE/RSJ International Conference on Intelligent Robots and Systems, IROS 2001: Expanding the Societal Role of Robotics in the the Next Millennium, Maui, HI, USA, October 29 - November 3, 2001*, pages 740–745.

Monica N. Nicolescu and Maja J. Mataric. 2003. Natural methods for robot task learning: instructive demonstrations, generalization and practice. In *The Second International Joint Conference on Autonomous Agents & Multiagent Systems, AAMAS 2003, July 14-18, 2003, Melbourne, Victoria, Australia, Proceedings*, pages 241–248.

JL. Part and O. Lemon. 2019. Towards a robot architecture for situated lifelong object learning. In *2019 IEEE/RSJ International Conference on Intelligent Robots and Systems (IROS)*, pages 1854–1860. IEEE.

Matthias Scheutz, Evan A. Krause, Bradley Oosterveld, Tyler M. Frasca, and Robert Platt. 2017. Spoken instruction-based one-shot object and action learning in a cognitive robotic architecture. In *AAMAS*.

Lanbo She and Joyce Yue Chai. 2017. Interactive learning of grounded verb semantics towards human-robot communication. In *ACL*.

Lanbo She, Shaohua Yang, Yu Cheng, Yunyi Jia, Joyce Yue Chai, and Ning Xi. 2014. Back to the blocks world: Learning new actions through situated human-robot dialogue. In *SIGDIAL Conference*.

G. Skantze. 2007. *Error handling in spoken dialogue systems: Managing uncertainty, grounding and miscommunication.* Gabriel Skantze.

Sida I. Wang, Percy Liang, and Christopher D. Manning. 2016. Learning language games through interaction. In *Proceedings of the 54th Annual Meeting of the Association for Computational Linguistics, ACL 2016, August 7-12, 2016, Berlin, Germany, Volume 1: Long Papers*.

Luke S. Zettlemoyer and Michael Collins. 2007. Online learning of relaxed ccg grammars for parsing to logical form. In *EMNLP-CoNLL*.

Learning to Read Maps: Understanding Natural Language Instructions from Unseen Maps

Miltiadis Marios Katsakioris[1], Ioannis Konstas[1], Pierre Yves Mignotte[2], Helen Hastie[1]
[1]School of Mathematical and Computer Sciences
Heriot-Watt University, Edinburgh, UK
[2]SeeByte Ltd, Edinburgh, UK
`Mmk11, I.konstas, H.hastie@hw.ac.uk`
`Pierre-yves.mignotte@seebyte.com`

Abstract

Robust situated dialog requires the ability to process instructions based on spatial information, which may or may not be available. We propose a model, based on LXMERT, that can extract spatial information from text instructions and attend to landmarks on OpenStreetMap (OSM) referred to in a natural language instruction. Whilst, OSM is a valuable resource, as with any open-sourced data, there is noise and variation in the names referred to on the map, as well as, variation in natural language instructions, hence the need for data-driven methods over rule-based systems. This paper demonstrates that the gold GPS location can be accurately predicted from the natural language instruction and metadata with 72% accuracy for previously seen maps and 64% for unseen maps.

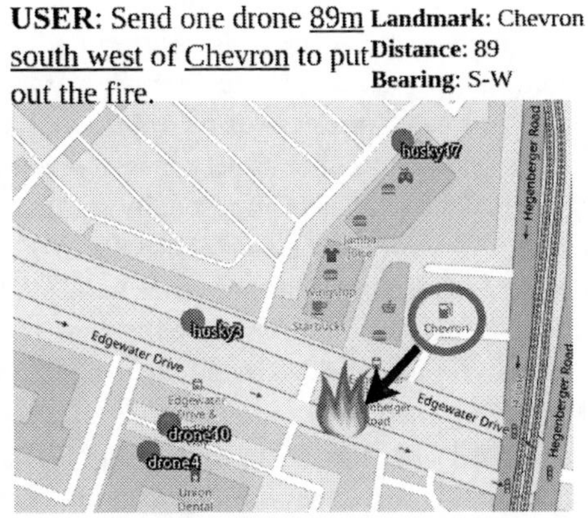

Figure 1: User instruction and the corresponding image, displaying 4 robots and landmarks. The users were not restricted or prompted to use specific landmarks on the map. The circle around the target landmark was added for clarity for this paper; users were not given any such visual hints.

1 Introduction

Spoken dialog systems are moving into real world situated dialog, such as assisting with emergency response and remote robot instruction that require knowledge of maps or building schemas. Effective communication of such an intelligent agent about events happening with respect to a map requires learning to associate natural language with the world representation found within the map. This symbol grounding problem (Harnad, 1990) has been largely studied in the context of mapping language to objects in a situated simple (MacMahon et al., 2006; Johnson et al., 2017) or 3D photorealistic environments (Kolve et al., 2017; Savva et al., 2019), static images (Ilinykh et al., 2019; Kazemzadeh et al., 2014), and to a lesser extent on synthetic (Thompson et al., 1993) and real geographic maps (Paz-Argaman and Tsarfaty, 2019; Haas and Riezler, 2016; Götze and Boye, 2016). The tasks usually relate to navigation (Misra et al., 2018; Thomason et al., 2019) or action execution (Bisk et al., 2018; Shridhar et al., 2019) and assume giving instructions to an embodied egocentric agent with a shared first-person view. Since most rely on the visual modality to ground natural language (NL), referring to items in the immediate surroundings, they are often less geared towards the accuracy of the final goal destination.

The task we address here is the prediction of the GPS of this goal destination by reference to a map, which is of critical importance in applications such as emergency response where specialized personnel or robots need to operate on an exact location (see Fig. 1 for an example). Specifically, the goal we are trying to predict is in terms of: a) the GPS coordinates (latitude/longitude) of a referenced landmark; b) a compass direction (bearing) from this referenced landmark; and c) the distance in meters from the referenced landmark. This is done by taking as input into a model: i) the knowledge base of the symbolic representation of the world such as landmark names and regions of interest (metadata); ii) the graphic depiction of a map

Proceedings of Second International Combined Workshop on Spatial Language Understanding andGrounded Communication for Robotics, pages 11–21
August 5–6, 2020.

(visual modality); and iii) a worded instruction.

Our approach to the destination prediction task is two-fold. The first stage is a data collection for the "Robot Open Street Map Instructions" (ROSMI) (Katsakioris et al., 2020) corpus based on OpenStreetMap (Haklay and Weber, 2008), in which we gather and align NL instructions to their corresponding target destination. We collected 560 NL instruction pairs on 7 maps of different variety and landmarks, in the domain of emergency response using Amazon Mechanical Turk. The subjects are given a scene in the form of a map and are tasked to write an instruction to command a conversational assistant to direct robots and autonomous systems to either inspect an area or extinguish a fire. The setup was intentionally emulating a typical *'Command and Control'* interface found in emergency response hubs, in order to promote instructions that accurately describe the final destination, with regards to its surrounding map entities.

Whilst OSM and other crowdsourced resources are hugely valuable, there is an element of noise associated with the metadata collected in terms of the names of the objects on the map, which can vary for the same type of object (e.g. newsagent/kiosk, confectionary/chocolate store etc.), whereas the symbols on the map are from a standard set, which one hypothesizes a vision-based trained model could pick-up on. To this end, we developed a model that leverages both vision and metadata to process the NL instructions.

Specifically, our MAPERT (Map Encoder Representations from Transformers) is a Transformer-based model based on LXMERT. It comprises of up to three single-modality encoders for each input (i.e., vision, metadata and language), an early fusion of modalities components and a cross-modality encoder, which fuses the map representation (metadata and/or vision) with the word embeddings of the instruction in both directions, in order to predict the three outputs, i.e., reference landmark location on the map, bearing and distance.

Our contributions are thus three-fold:

- A novel task for final GPS destination prediction from NL instructions with accompanying ROSMI dataset[1].
- A model that predicts GPS goal locations from a map-based natural language instruction.
- A model that is able to understand instructions referring to previously unseen maps.

2 Related Work

Situated dialog encompasses various aspects of interaction. These include: situated Natural Language Processing (Bastianelli et al., 2016); situated reference resolution (Misu, 2018); language grounding (Johnson et al., 2017); visual question answer/visual dialog (Antol et al., 2015); dialog agents for learning visually grounded word meanings and learning from demonstration (Yu et al., 2017); and Natural Language Generation (NLG), e.g. of situated instructions and referring expressions (Byron et al., 2009; Kelleher and Kruijff, 2006). Here, work on instruction processing for destination mapping and navigation are discussed, as well as language grounding and referring expression resolution, with an emphasis on 2D/3D real world and map-based application.

Language grounding refers to interpreting language in a situated context and includes collaborative language grounding toward situated human-robot dialog (Chai et al., 2016), city exploration (Boye et al., 2014), as well as following high-level navigation instructions (Blukis et al., 2018). Mapping instructions to low level actions has been explored in structured environments by mapping raw visual representations of the world and text onto actions using using Reinforcement Learning methods (Misra et al., 2017; Xiong et al., 2018; Huang et al., 2019). This work has recently been extended to controlling autonomous systems and robots through human language instruction in a 3D simulated environment (Ma et al., 2019; Misra et al., 2018; Blukis et al., 2019) and Mixed Reality (Huang et al., 2019) and using imitation learning (Blukis et al., 2018). These systems perform goal prediction and action generation to control a single Unmanned Aerial Vehicles (UAVs), given a natural language instruction, a world representation and/or robot observations. However, where this prior work uses raw pixels to generate a persistent semantic map from the system's line-of-sight image, our model is able to leverage both pixel and metadata, when it is available in a combined approach. Other approaches include neural mapping of navigational instructions to action sequences (Mei et al., 2015), which does include a representation of the observable world state, but this is more akin to a maze rather than a complex map.

[1]We make our code and data available at https://github.com/marioskatsak/mapert.

With respect to the task, our model looks to predict GPS locations. There are few related works that attempt this challenging task. One study, as part of the ECML/PKDD challenge (de Brébisson et al., 2015), uses Neural Networks for Taxi Destination Prediction as a sequence of GPS points. However, this does not include processing natural language instructions. SPACEREF (Götze and Boye, 2016) is perhaps the closest to our task in that the task entails both GPS tracks in OSM and annotated mentions of spatial entities in natural language. However, it is different in that these spatial entities are viewed and referred to in a first person view, rather than entities on a map (e.g. "the arch at the bottom").

In terms of our choice of model, attention mechanisms (Bahdanau et al., 2015; Vaswani et al., 2017; Xu et al., 2015) have proven to be very powerful in language and vision tasks and we draw inspiration from the way (Xu et al., 2015) use attention to solve image captioning by associating words to spatial regions within a given image.

3 Data

As mentioned above, the task is based on OpenStreetMap (OSM) (Haklay and Weber, 2008). OSM is a massively collaborative project, started in 2004, with the main goal to create a free editable map of the world. The data is available under the Open Data Commons Open Database Licence and has been used in some prior work (Götze and Boye, 2016; Hentschel and Wagner, 2010; Haklay and Weber, 2008). It is a collection of publicly available geodata that are constantly updated by the public and consists of many layers of various geographic attributes of the world. Physical features such as roads or buildings are represented using tags (metadata) that are attached to its basic data structures. A comprehensive list of all the possible features available as metadata can be found online[2]. There are two types of objects, *nodes* and *ways*, with unique IDs that are described by their latitude/longitude (lat/lon) coordinates. Nodes are single points (e.g. coffee shops) whereas ways can be more complex structures, such as polygons or lines (e.g. streets and rivers). For this study, we train and test only on data that uses single points (nodes) and polygons (using the centre point), and leave understanding more complex structures as future work.

We train and evaluate our model on ROSMI, a new multimodal corpus. This corpus consists of visual and natural language instruction pairs, in the domain of emergency response. In this data collection, the subjects were given a scene in the form of an OSM map and were tasked to write an instruction to command a conversational assistant to direct a number of robots and autonomous systems to either inspect an area or extinguish a fire. Figure 1 shows an example of such a written instruction. These types of emergency scenarios usually have a central hub for operators to observe and command humans and Robots and Autonomous Systems (RAS) to perform specific functions, where the robotic assets are visually observable as an overlay on top of the map. Each instruction datapoint was manually checked and if it did not match the 'gold standard' GPS coordinate per the scenario map, it was discarded. The corpus was manually annotated with the ground truth for, (1) a link between the NL instruction and the referenced OSM entities; and (2) the distance and bearing from this referenced entity to the goal destination. The ROSMI corpus thus comprises 560 tuples of instructions, maps with metadata and target GPS location.

There are three linguistic phenomena of note that we observe in the data collected. Firstly, **Landmark Grounding** where each scenario has 3-5 generated *robots* and an average of 30 *landmarks* taken from OSM. Each subject could refer to any of these objects on the map, in order to complete the task. Grounding the right noun phrase to the right OSM landmark or robot, is crucial for predicting accurately the gold-standard coordinate, e.g. *send husky11 62m to the west direction* or *send 2 drones near Harborside Park.*

Secondly, **Bearing/Distance** factors need to be extracted from the instruction such as numbers (e.g. 500 meters) and directions (e.g. northwest, NE) and these two items typically come together. For example, *"send drone11 to the west about 88m"*.

Thirdly, **Spatial Relations** are where prepositions are used instead of distance/bearing (e.g. near, between), and are thus more vague. For example, *"Send a drone near the Silver Strand Preserve"*.

4 Approach

4.1 Task Formulation

An instruction is taken as a sequence of word tokens $\mathbf{w} =< w_1, w_2, \ldots w_N >$ with $\mathrm{w_i} \in V$,

[2] wiki.openstreetmap.org/wiki/Map_Features

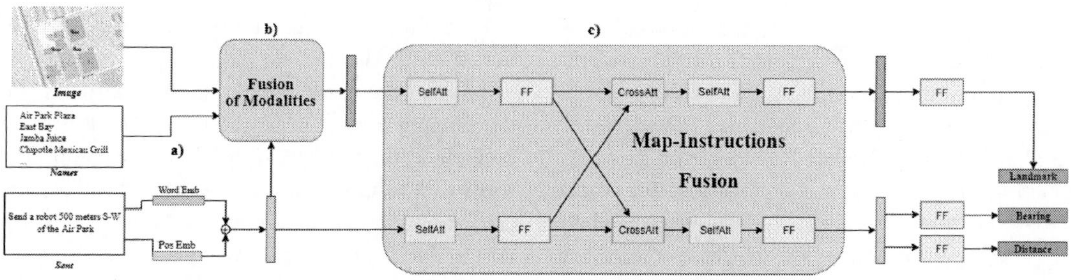

Figure 2: Architecture of MAPERT. Map representations, i.e., names of landmarks found in OSM (metadata) and Faster-RCNN predicted objects (visual modality), along with an instruction (sequence of tokens) are a) encoded into the model, b) fused together (see also Fig. 4) and c) bidirectionally attended. The output comprises of three predictions, recast as classification tasks: a landmark, a bearing and a distance.

where V is a vocabulary of words and the corresponding geographic map I is represented as a set of M landmark objects $o_i = (\mathbf{bb}, \mathbf{r}, \mathbf{n})$ where $\mathbf{bb}$ is a 4-dimensional vector with bounding box coordinates, $\mathbf{r}$ is the corresponding Region of Interest (RoI) feature vector produced by an object detector and $n =< n_1, n_2 \ldots n_K >$, is a multi-token name. We define a function $f : V^N \times R^{4*M} \times R^{2048*M} \times V^{M*K} \to R \times R$ to predict the GPS destination location $\hat{y}$:

$$\hat{y} = f\big(\mathbf{w}, \{o_i = (\mathbf{bb}, \mathbf{r}, \mathbf{n})\}_M\big) \quad (1)$$

Since predicting $\hat{y}$ directly from $\mathbf{w}$ is a harder task, we decompose it into three simpler components, namely predicting a reference *landmark* location $l \in M$, the compass direction (bearing) b[3], and a distance d from l in meters. Then we trivially convert to the final GPS position coordinates. Equation 1 now becomes:

$$\hat{y} = gps(l, d, b) = f\big(\mathbf{w}, \{o_i = (\mathbf{bb}, \mathbf{r}, \mathbf{n})\}_M\big) \quad (2)$$

4.2 Model Architecture

Inspired by LXMERT (Tan and Bansal, 2019), we present MAPERT, a Transformer-based (Vaswani et al., 2017) model with three separate single-modality encoders (for NL instructions, metadata and visual features) and a cross-modality encoder that merges them. Fig. 2 depicts the architecture. In the following sections, we describe each component separately.

Instructions Encoder The word sequence $\mathbf{w}$ is fed to a Transformer encoder and output hidden states $\mathbf{h_w}$ and position embeddings $\mathbf{pos_w}$; its weights are initialized using pretrained BERT (Devlin et al., 2019). $\mathbf{h_{w_0}}$ is the hidden state for the special token [CLS].

Metadata Encoder OSM comes with useful metadata in the form of bounding boxes (around the landmark symbols) and names of landmarks on the map. We represent each bounding box as a 4-dimensional vector $\mathbf{bb_{meta_k}}$ and each name ($\mathbf{n_k}$) using another Transformer initialized with pretrained BERT weights. We treat metadata as a bag of names but since each word can have multiple tokens, we output position embeddings $\mathbf{pos_{n_k}}$ for each name separately; $\mathbf{h_{n_k}}$ are the resulting hidden states with $\mathbf{h_{n_{k,0}}}$ being the hidden state for [CLS].

Visual Encoder Each map image is fed into a pretrained Faster R-CNN detector (Ren et al., 2015), which outputs bounding boxes and RoI feature vectors $\mathbf{bb_k}$ and $\mathbf{r_k}$ for k objects. In order to learn better representation for landmarks, we fine-tuned the detector on around 27k images of maps to recognize k objects $\{o_1, .., o_k\}$ and classify landmarks of 213 manually-cleaned classes from OSM; we fixed k to 73 landmarks. Finally, a combined position-aware embedding $\mathbf{v_k}$ was learned by adding together the vectors $\mathbf{bb_k}$ and $\mathbf{r_k}$ as in LXMERT:

$$\mathbf{v_k} = \frac{FF(\mathbf{bb_k}) + FF(\mathbf{r_k})}{2} \quad (3)$$

where FF are feed-forward layers with no bias.

4.3 Variants for Fusion of Input Modalities

We describe three different approaches to combining knowledge from maps with the NL instructions:

Metadata and Language The outputs of the metadata and language encoders are fused by conditioning each landmark name n_i on the instruction

[3] $b \in \{N, NE, NW, E, SE, S, SW, SE, W, None\}$.

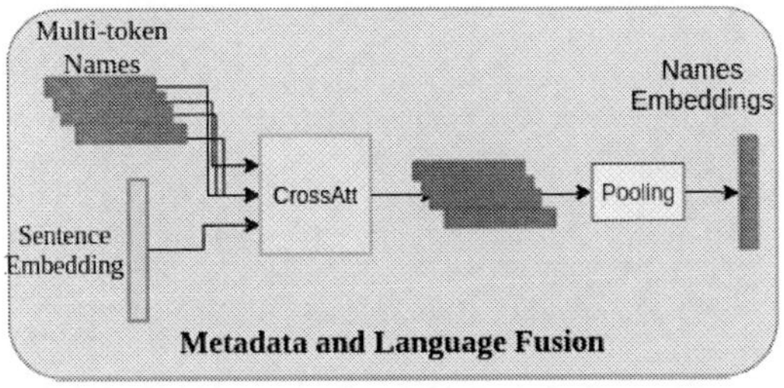

Figure 3: Metadata and Language fusion module. Multi-token names correspond to the BERT-based embeddings of landmarks names. The output is the embedding used to represent the landmarks names from OSM metadata.

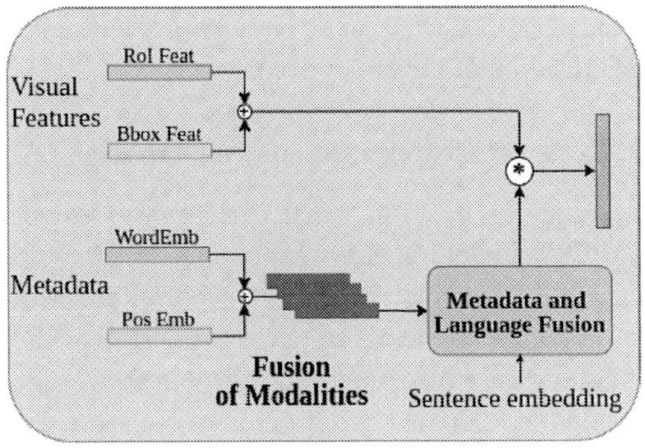

Figure 4: Fusion of metadata, vision and language modalities. Metadata are first conditioned on the instruction tokens as shown in Fig. 3. Then, they are multiplied with the visual features of every landmark.

sequence via a uni-directional cross attention layer (Fig. 3). We first compute the attention weights A_k between the name tokens $\mathbf{n_{k,i}}$ of each landmark o_k and instruction words in $\mathbf{h_w}$[4] and re-weight the hidden states $\mathbf{h_{n_k}}$ to get the context vectors $\mathbf{c_{n_k}}$. We then pool them using the context vector for the [CLS] token of each name:

$$\mathbf{A_k} = CrossAttn(\mathbf{h_w}, \mathbf{n_k}) \quad (4)$$

$$\mathbf{c_{n_k}} = \mathbf{A_k} \odot \mathbf{n_k} \quad (5)$$

$$\mathbf{h_{meta}} = BertPooler(\mathbf{c_{n_k}}) \quad (6)$$

We can also concatenate the bounding box $\mathbf{bb_{meta_k}}$ to the final hidden states:

$$\mathbf{h_{meta+bb}} = [\mathbf{h_{meta}}; FF(\mathbf{bb_{meta_k}})] \quad (7)$$

Metadata+Vision and Language All three modalities were fused to verify whether vision can aid metadata information for the final GPS destination prediction task (Fig. 4). First, we filter the landmarks o_i based on the Intersection over Union between the bounding boxes found in metadata ($\mathbf{bb_{meta_k}}$) and those predicted with Faster R-CNN ($\mathbf{bb_k}$), thus keeping their corresponding names n_i and visual features $\mathbf{v_i}$. Then, we compute the instruction-conditioned metadata hidden states $\mathbf{h_{meta_i}}$, as described above, and multiply them with every object v_i to get the final $\mathbf{h_{meta+vis}}$ context vectors:

$$\mathbf{h_{meta+vis_i}} = \mathbf{h_{meta_i}} \otimes \mathbf{v_i} \quad (8)$$

[4]Whenever we refer to hidden states $\mathbf{h_w}$ we assume concatenation with corresponding positional embeddings $[\mathbf{h_w};\mathbf{pos_w}]$, which we omit here for brevity.

4.4 Map-Instructions Fusion

So far we have conditioned modalities in one direction, i.e., from the instruction to metadata and visual features. In order to capture the influence between map and instructions in both ways, a cross-modality encoder was implemented (right half of Fig. 2). Firstly each modality passes through a self-attention and feed-forward layer to highlight inter-dependencies. Then these modulated inputs are passed to the actual fusion component, which consists of one bi-directional cross-attention layer, two self-attention layers, and two feed-forward layers. The cross-attention layer is a combination of two unidirectional cross-attention layers, one from instruction tokens ($\mathbf{h_w}$) to map representations (either of $\mathbf{h_{meta_k}}$, $\mathbf{v_k}$ or $\mathbf{h_{meta+vis_k}}$; we refer to them below as $\mathbf{h_{map_k}}$) and vice-versa:

$$\tilde{\mathbf{h}}_\mathbf{w} = FF(SelfAtt(\mathbf{h_w})) \quad (9)$$

$$\tilde{\mathbf{h}}_\mathbf{map_k} = FF(SelfAtt(\mathbf{h_{map_k}})) \quad (10)$$

$$\mathbf{C_{map_k}} = CrossAtt(\tilde{\mathbf{h}}_\mathbf{w}, \tilde{\mathbf{h}}_\mathbf{map_k}) \quad (11)$$

$$\mathbf{C_w} = CrossAtt(\tilde{\mathbf{h}}_\mathbf{map_k}, \tilde{\mathbf{h}}_\mathbf{w}) \quad (12)$$

$$\mathbf{h_{cross,w}} = \mathbf{C_w} \odot \tilde{\mathbf{h}}_\mathbf{w} \quad (13)$$

$$\mathbf{h_{cross,map_k}} = \mathbf{C_{map_k}} \odot \tilde{\mathbf{h}}_\mathbf{map_k} \quad (14)$$

$$\mathbf{out_w} = FF(SelfAtt(\mathbf{h_{cross,w}})) \quad (15)$$

$$\mathbf{out_{map_k}} = FF(SelfAtt(\mathbf{h_{cross,map_k}})) \quad (16)$$

Note that representing $\mathbf{h_{map_k}}$ with vision features $\mathbf{v_k}$ only is essentially a fusion between the vision and language modalities. This is a useful variant of our model to measure whether the visual representation of a map alone is as powerful as metadata, specifically for accurately predicting the GPS location of the target destination.

4.5 Output Representations and Training

As shown in the right-most part of Fig. 2, our MAPERT model has three outputs: landmarks, distances, and bearings. We treat each output as a classification sub-task, i.e., predicting one or the k landmarks in the map; identifying in the NL instruction the start and end position of the sequence of tokens that denotes a distance from the reference landmark (e.g., *'500m'*); and a bearing label. MAPERT's output comprises of two feature vectors, one for the vision and one for the language modality generated by the cross-modality encoder.

More specifically, for the bearing predictor, we pass the hidden state $\mathbf{out_{w,0}}$, corresponding to [CLS], to a FF followed by a softmax layer. Predicting distance is similar to span prediction for Question Answering tasks; we project each of the tokens in $\mathbf{out_w}$ down to 2 dimensions corresponding to the distance span boundaries in the instruction sentence. If there is no distance in the sentence e.g., *"Send a drone at Jamba Juice"*, the model learns to predict, both as start and end position, the final end of sentence symbol, as an indication of absence of distance. Finally, for landmark prediction we project each of the k map hidden states $\mathbf{out_{map_k}}$ to a single dimension corresponding to the index of the ith landmark.

We optimize MAPERT by summing the cross-entropy losses for each of the classification sub-tasks. The final training objective becomes:

$$\mathcal{L} = \mathcal{L}_{land} + \mathcal{L}_{bear} + \mathcal{L}_{dist,start} + \mathcal{L}_{dist,end} \quad (17)$$

5 Experimental Setup

Implementation Details We evaluate our model on the ROSMI dataset and assess the contribution of the metadata and vision components as described above. For the attention modules, we use a hidden layer with size of 768 as in $BERT_{BASE}$ and we set the numbers of all the encoder and fusion layers to 1. We initialize pretrained BERT embedding layers (we also show results with randomly initialized embeddings). We trained our model using Adam (Kingma and Ba, 2015) as the optimizer with a linear-decayed learning-rate schedule (Tan and Bansal, 2019) for 90 epochs, a dropout probability of 0.1 and learning rate of 10^{-3}.

	10-fold Cross Validation (unseen examples)	
	$\text{Acc}_{50}[SD]$	T Err(m) [SD]
Oracle_{lower}	80 [5.01]	23.8 [51.9]
Vision		
bbox	46.18 [5.59]	44.7 [51.7]
RoI+bbox	60.36 [5.3]	36.4 [51.1]
Meta+Vision		
RoI+bbox+names	69.27 [6.68]	26.9 [47.7]
Meta		
bbox	46.18 [5.59]	44.7 [51.7]
names	**71.81 [7.37]**	26.7 [47.7]
bbox+names	70.73 [6.58]	26.3 [48.7]
Oracle_{upper}	100 [0.0]	0 [0]
Meta		
bbox	60.36 [5.26]	29.8 [44.9]
names	**87.64 [4.8]**	9.6 [29.9]
bbox+names	87.09 [5.66]	9.5 [27.2]

Table 1: Ablation results on ROSMI using a 10-fold cross validation. Accuracy (Acc) with IoU of 0.5 and Targer error (T Err) in meters. The results in the top half of the table use names conditioned on the lower bound of the Vision modality and so are compared to Oracle_{lower}. The bottom part of the table use the true metadata names and so are to be compared to Oracle_{upper}.

Evaluation Metrics We use a 10-fold cross-validation for our evaluation methodology. This results in a less biased estimate of the accuracy over splitting the data into train/test due to the modest size of the dataset. In addition, we performed a leave-one-map-out cross-validation, as in Chen and Mooney (2011). In other words, we use 7-fold cross-validation, and in each fold we use six maps for training and one map for validation. We refer to these scenarios as zero-shot[5] since, in each fold, we validate our data on an unseen map scenario. With the three outputs of our model, landmark, distance and bearing, we indirectly predict the destination location. Success is measured by the Intersection over Union (IoU) between the ground truth destination location and the calculated destination location. IoU measures the overlap between two bounding boxes and as in Everingham et al. (2010), must exceed 0.5 (50%) to count it as successful by the formula:

$$IoU = \frac{area(B_p \cap B_{gt})}{area(B_p \cup B_{gt})} \quad (18)$$

Since we are dealing with GPS coordinates but also image pixels, we report two error evaluation

[5] We loosely use the term zero-shot as we appreciate that there might be some overlap in terms of street names and some objects

metrics. The first is sized weighted Target error (T err) in meters, which is the distance in meters between the predicted GPS coordinate and the ground truth coordinate. The second is a Pixel Error (P error) which is the difference in pixels between the predicted point in the image and the ground truth converted from the GPS coordinate.

Comparison of Systems We evaluate our system on three variants using different fusion techniques, namely Meta and Language; Meta+Vision and Language; and Vision and Language. Ablations for these systems are shown in Table 1 and are further analyzed in Section 6. We also compare MAPERT to a strong baseline, BERT. The baseline is essentially MAPERT but without the bidirectional cross attention layers in the pipeline (see Fig. 2).

Note, the Oracle of the Meta and Language has a 100% (upper bound) on both cross-validation splits of ROSMI, whereas the oracle of any model that utilizes visual features, is 80% in the 10-fold and 81.98% in the 7-fold cross-validation (lower bound). In other words, the GPS predictor can only work with the output of the automatically predicted entities outputed from Faster R-CNN, of which 20% are inaccurate. Table 1 shows results on both oracles, with the subscript *lower* indicating the lower bound oracle and *upper* indicating the "Upper Bound" oracle. In Table 2, all systems are being projected on the lower bound oracle, so as to compare them on the same footing.

6 Results

Table 2 shows the results of our model for Vision, Meta and Meta+Vision on both the 10-fold cross validation and the 7-fold zero-shot cross validation. We see that the Meta variant of MAPERT outperforms all other variants and our baseline. However, looking at the 10-fold results, Meta+Vision's accuracy of 69.27% comes almost on par with Meta's 71.81%. If we have the harder task of no metadata, with only the visuals of the map to work with, we can see that the Vision component works reasonably well, with an accuracy to 60.36%. This Vision component, despite being on a disadvantage, manages to learn the relationship of visual features with an instruction and vice-versa, compared to our baseline, which has no crossing between the modalities whatsoever, reaching only 33.82%. When we compare these results to the zero-shot paradigm, we see only a 10.5% reduction using Meta, whereas

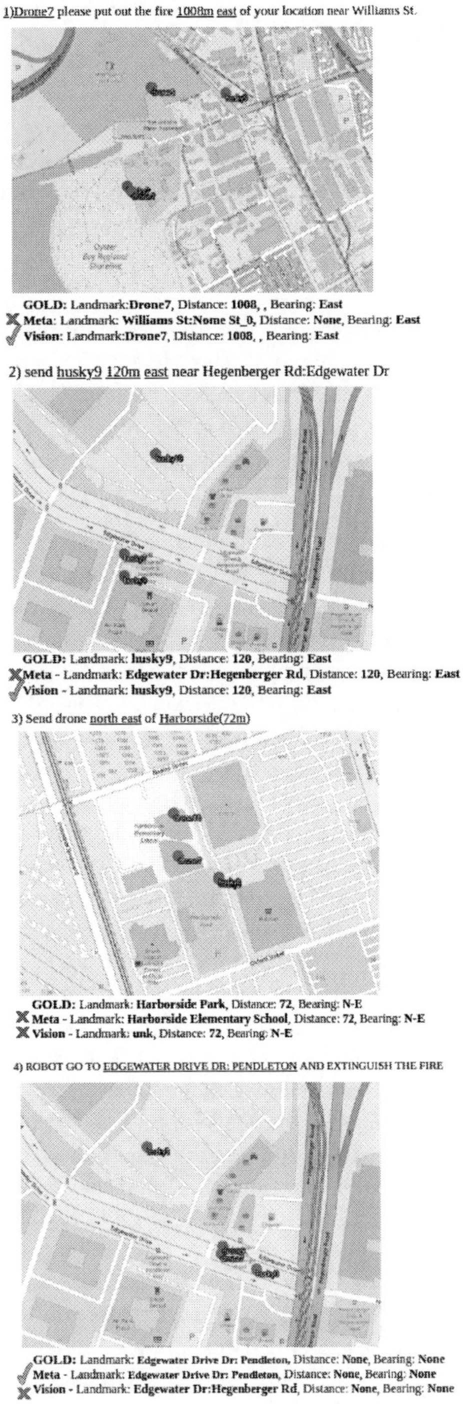

Figure 5: Examples of instructions with the corresponding maps and the accompanied predictions of the best performing either Vision or Meta models conditioned on Oracle$_{lower}$. Underlined words are words corresponding to the target output of the model.

	10-fold Cross Validation (unseen examples)			7-fold Cross Validation (unseen scenarios)		
	Accuracy$_{50}$ [SD]	T err [SD]	P err [SD]	Accuracy$_{50}$ [SD]	T err (m)[SD]	P err (m) [SD]
Oracle$_{lower}$	80 [5.01]	23.8 [51.9]	39.1 [96.3]	81.98 [17.09]	20.14 [39]	33.29 [66.43]
Baseline	33.82 [5.16]	64 [57.1]	119.8 [112.3]	34.90 [11.13]	60.71 [57.14]	110.43 [109.71]
Meta	**71.81 [7.37]**	26.70 [47.7]	48.2 [91.2]	**64.30 [14.16]**	32.71 [50.14]	65.71 [88.4]
Vision	60.36 [5.30]	36.40 [51.1]	64.40 [99.6]	49.75 [8.06]	46.00 [54.57]	87.86 [106.0]
Meta+Vision	69.27 [6.68]	26.90 [47.7]	48.30 [91.4]	58.33 [12.24]	36.14 [46.14]	70.71 [93.29]

Table 2: Results on both cross-validations of the best performing ablations of each variant and the baseline. The predictions have been made under the Oracle$_{lower}$. Accuracy (Acc) with IoU of 0.5, Target error (T Err) and Pixel Error (P Err) in meters.

the Vision only component struggles more, with a 17.6% reduction and Vision+Meta a 15.8% reduction. This is understandable since on the 7-fold validation, we tackle unseen maps, which is very challenging for the Vision-only model.

Ablation Study We show ablations for all three model variants in Table 1 and corresponding ablations. We show here just the 10-fold as the 7-fold has similar performance ordering. Depending on the representation of the map for each variant, we derive three ablations for the Meta and two for the Vision. Meta+Vision does not have ablations, since it stands for all possible representations (bb, r, n). Compared to the Oracle$_{lower}$, Meta outperforms the rest, as seen in Table 2. In addition, it requires only the names of the landmarks to score the 71.73%. When we fuse the names and the bboxes, the accuracy decreases slightly, whereas the T err decreases slightly from 26.7 meters to 26.3 meters. The full potential of the Meta model is shown on the Oracle$_{upper}$, which reaches 87.64 % accuracy and T Err of only 9.6 meters, proof that for our task and dataset metadata has the upper hand. It is worthwhile noting that the Vision variant would not have reached 60.36% accuracy, without the r features, since with no fusion of RoI, the accuracy drops to 46.18%.

Error Analysis In order to understand where the Vision and Meta models' comparative strengths lie, we show some example outputs in Fig. 5. In examples 1&2 in this figure, we see the Meta model is failing to identify the correct landmark because the instruction is formulated in a way that allows the identification of two landmarks. It's a matter of which landmark to choose, and the bearing, distance that comes with it, to successfully predict the destination location. However, the Meta model is mixing up the landmarks and the bearings. We believe it is that perhaps the Meta model struggles with spatial relations such as "near". The Vision model, on the other hand, successfully picks up the three correct components for the prediction. This might be helped by the familiarity of the symbolic representation the robots (husky, drones, auvs), which it is able to pick up and use as landmarks in situations of uncertainty such as this one. Both models can fail in situations of both visual and metadata ambiguity. In the third example, the landmark (Harborside Park) is not properly specified and both models fail to pinpoint the correct landmark, since further clarification would be needed. The final example in Fig. 5 shows a situation in which the Meta model works well without the need of a specific distance and bearing. The Vision model manages to capture that, but it fails to identify the correct landmark.

7 Conclusion and Future Work

We have developed a model that is able to process instructions on a map using metadata from rich map resources such as OSM and can do so for maps that it has not seen before with only a 10% reduction in accuracy. If no metadata is available then the model can use Vision, although this is clearly a harder task. Vision does seem to help in examples where there is a level of uncertainty such as with spatial relations or ambiguity between entities. Future work will involve exploring this further by training the model on these type of instructions and on metadata that are scarce and inaccurate. Finally, these instructions will be used in an end-to-end dialog system for remote robot planning, whereby multi-turn interaction can handle ambiguity and ensure reliable and safe destination prediction before instructing remote operations.

Acknowledgments

Katsakioris' PhD was funded by SeeByte, Datalab and MASTS-S. This work was also supported by the EPSRC funded ORCA Hub (EP/R026173/1, 2017-2021).

References

Stanislaw Antol, Aishwarya Agrawal, Jiasen Lu, Margaret Mitchell, Dhruv Batra, C. Lawrence Zitnick, and Devi Parikh. 2015. VQA: Visual Question Answering. In *International Conference on Computer Vision (ICCV)*.

Dzmitry Bahdanau, Kyunghyun Cho, and Yoshua Bengio. 2015. Neural machine translation by jointly learning to align and translate. In *3rd International Conference on Learning Representations, ICLR 2015, San Diego, CA, USA, May 7-9, 2015, Conference Track Proceedings*.

Emanuele Bastianelli, Danilo Croce, Andrea Vanzo, Roberto Basili, and Daniele Nardi. 2016. A discriminative approach to grounded spoken language understanding in interactive robotics. In *Proceedings of the Twenty-Fifth International Joint Conference on Artificial Intelligence*, IJCAI'16, pages 2747–2753. AAAI Press.

Yonatan Bisk, Kevin Shih, Yejin Choi, and Daniel Marcu. 2018. Learning interpretable spatial operations in a rich 3d blocks world. In *Proceedings of the Thirty-Second Conference on Artificial Intelligence (AAAI-18)*.

Valts Blukis, Nataly Brukhim, Andrew Bennett, Ross A. Knepper, and Yoav Artzi. 2018. Following high-level navigation instructions on a simulated quadcopter with imitation learning. In *Robotics: Science and Systems (RSS)*.

Valts Blukis, Yannick Terme, Eyvind Niklasson, Ross A. Knepper, and Yoav Artzi. 2019. Learning to map natural language instructions to physical quadcopter control using simulated flight. In *Proceedings of the Conference on Robot Learning*.

Johan Boye, Morgan Fredriksson, Jana Götze, Joakim Gustafson, and Jürgen Königsmann. 2014. Walk this way: Spatial grounding for city exploration. In *Natural Interaction with Robots, Knowbots and Smartphones*, pages 59–67, New York, NY. Springer New York.

Alexandre de Brébisson, Étienne Simon, Alex Auvolat, Pascal Vincent, and Yoshua Bengio. 2015. Artificial neural networks applied to taxi destination prediction. *CoRR*, abs/1508.00021.

Donna Byron, Alexander Koller, Kristina Striegnitz, Justine Cassell, Robert Dale, Johanna Moore, and Jon Oberlander. 2009. Report on the First NLG Challenge on Generating Instructions in Virtual Environments (GIVE). In *Proceedings of the 12th European Workshop on Natural Language Generation (ENLG 2009)*, pages 165–173, Athens, Greece. Association for Computational Linguistics.

Joyce Yue Chai, Rui Fang, Changsong Liu, and Lanbo She. 2016. Collaborative language grounding toward situated human-robot dialogue. *AI Magazine*, 37:32–45.

David L. Chen and Raymond J. Mooney. 2011. Learning to interpret natural language navigation instructions from observations. In *Proceedings of the Twenty-Fifth AAAI Conference on Artificial Intelligence*. AAAI Press.

Jacob Devlin, Ming-Wei Chang, Kenton Lee, and Kristina Toutanova. 2019. BERT: Pre-training of deep bidirectional transformers for language understanding. In *Proceedings of the 2019 Conference of the North American Chapter of the Association for Computational Linguistics: Human Language Technologies*, Minneapolis, Minnesota. Association for Computational Linguistics.

Mark Everingham, Luc Gool, Christopher K. Williams, John Winn, and Andrew Zisserman. 2010. The pascal visual object classes (VOC) challenge. *Int. J. Comput. Vision*, 88(2):303–338.

Jana Götze and Johan Boye. 2016. SpaceRef: A corpus of street-level geographic descriptions. In *Proceedings of the Tenth International Conference on Language Resources and Evaluation (LREC 2016)*, pages 3822–3827, Portorož, Slovenia. European Language Resources Association (ELRA).

Carolin Haas and Stefan Riezler. 2016. A corpus and semantic parser for multilingual natural language querying of OpenStreetMap. In *Proceedings of the 2016 Conference of the North American Chapter of the Association for Computational Linguistics: Human Language Technologies*, pages 740–750, San Diego, California. Association for Computational Linguistics.

M. Haklay and P. Weber. 2008. Openstreetmap: User-generated street maps. *IEEE Pervasive Computing*, 7(4):12–18.

Stevan Harnad. 1990. The symbol grounding problem. *Phys. D*, 42(1â3):335–346.

M. Hentschel and B. Wagner. 2010. Autonomous robot navigation based on openstreetmap geodata. In *Proceedings of the 13th International IEEE Conference on Intelligent Transportation Systems*, pages 1645–1650.

Baichuan Huang, Deniz Bayazit, Daniel Ullman, Nakul Gopalan, and Stefanie Tellex. 2019. Flight, Camera, Action! Using Natural Language and Mixed Reality to Control a Drone. In *IEEE International Conference on Robotics and Automation (ICRA)*.

Nikolai Ilinykh, Sina Zarrieß, and David Schlangen. 2019. Tell me more: A dataset of visual scene description sequences. In *Proceedings of the 12th International Conference on Natural Language Generation*, pages 152–157, Tokyo, Japan. Association for Computational Linguistics.

J. Johnson, B. Hariharan, L. van der Maaten, L. Fei-Fei, C. L. Zitnick, and R. Girshick. 2017. CLEVR: A diagnostic dataset for compositional language and elementary visual reasoning. In *2017 IEEE Conference on Computer Vision and Pattern Recognition (CVPR)*, pages 1988–1997.

Miltiadis Marios Katsakioris, Ioannis Konstas, Pierre Yves Mignotte, and Helen Hastie. 2020. Rosmi: A multimodal corpus for map-based instruction-giving. In *Proceedings of the 2020 International Conference on Multimodal Interaction*, ICMI '20, pages 680–684, New York, NY, USA. Association for Computing Machinery.

Sahar Kazemzadeh, Vicente Ordonez, Mark Matten, and Tamara L. Berg. 2014. Referit game: Referring to objects in photographs of natural scenes. In *Proceedings of EMNLP*.

John D. Kelleher and Geert-Jan M. Kruijff. 2006. Incremental generation of spatial referring expressions in situated dialog. In *Proceedings of the 21st International Conference on Computational Linguistics and the 44th Annual Meeting of the Association for Computational Linguistics*, ACL-44, pages 1041–1048, Stroudsburg, PA, USA. Association for Computational Linguistics.

Diederik P. Kingma and Jimmy Ba. 2015. Adam: A method for stochastic optimization. In *3rd International Conference on Learning Representations, ICLR 2015, San Diego, CA, USA, May 7-9, 2015, Conference Track Proceedings*.

Eric Kolve, Roozbeh Mottaghi, Daniel Gordon, Yuke Zhu, Abhinav Gupta, and Ali Farhadi. 2017. AI2-THOR: an interactive 3d environment for visual AI. *CoRR*, abs/1712.05474.

Chih-Yao Ma, Jiasen Lu, Zuxuan Wu, Ghassan AlRegib, Zsolt Kira, Richard Socher, and Caiming Xiong. 2019. Self-monitoring navigation agent via auxiliary progress estimation. *CoRR*, abs/1901.03035.

Matt MacMahon, Brian Stankiewicz, and Benjamin Kuipers. 2006. Walk the talk: Connecting language, knowledge, and action in route instructions. In *Proceedings of the 21st National Conference on Artificial Intelligence - Volume 2*, AAAI'06, pages 1475–1482. AAAI Press.

Hongyuan Mei, Mohit Bansal, and Matthew R. Walter. 2015. Listen, attend, and walk: Neural mapping of navigational instructions to action sequences. *CoRR*, abs/1506.04089.

Dipendra Misra, Andrew Bennett, Valts Blukis, Eyvind Niklasson, Max Shatkhin, and Yoav Artzi. 2018. Mapping instructions to actions in 3D environments with visual goal prediction. In *Proceedings of the 2018 Conference on Empirical Methods in Natural Language Processing*, pages 2667–2678, Brussels, Belgium. Association for Computational Linguistics.

Dipendra Misra, John Langford, and Yoav Artzi. 2017. Mapping instructions and visual observations to actions with reinforcement learning. In *Proceedings of the 2017 Conference on Empirical Methods in Natural Language Processing*, pages 1004–1015, Copenhagen, Denmark. Association for Computational Linguistics.

Teruhisa Misu. 2018. Situated reference resolution using visual saliency and crowdsourcing-based priors for a spoken dialog system within vehicles. *Computer Speech and Language*, 48:1 – 14.

Tzuf Paz-Argaman and Reut Tsarfaty. 2019. RUN through the streets: A new dataset and baseline models for realistic urban navigation. In *Proceedings of the 2019 Conference on Empirical Methods in Natural Language Processing and the 9th International Joint Conference on Natural Language Processing (EMNLP-IJCNLP)*, pages 6449–6455, Hong Kong, China. Association for Computational Linguistics.

Shaoqing Ren, Kaiming He, Ross Girshick, and Jian Sun. 2015. Faster r-cnn: Towards real-time object detection with region proposal networks. In C. Cortes, N. D. Lawrence, D. D. Lee, M. Sugiyama, and R. Garnett, editors, *Advances in Neural Information Processing Systems 28*, pages 91–99. Curran Associates, Inc.

Manolis Savva, Abhishek Kadian, Oleksandr Maksymets, Yili Zhao, Erik Wijmans, Bhavana Jain, Julian Straub, Jia Liu, Vladlen Koltun, Jitendra Malik, Devi Parikh, and Dhruv Batra. 2019. Habitat: A platform for embodied AI research. *CoRR*, abs/1904.01201.

Mohit Shridhar, Jesse Thomason, Daniel Gordon, Yonatan Bisk, Winson Han, Roozbeh Mottaghi, Luke Zettlemoyer, and Dieter Fox. 2019. ALFRED: A Benchmark for Interpreting Grounded Instructions for Everyday Tasks.

Hao Tan and Mohit Bansal. 2019. LXMERT: Learning cross-modality encoder representations from transformers. In *Proceedings of EMNLP-IJCNLP*, pages 5100–5111, Hong Kong, China. Association for Computational Linguistics.

Jesse Thomason, Michael Murray, Maya Cakmak, and Luke Zettlemoyer. 2019. Vision-and-dialog navigation. *CoRR*, abs/1907.04957.

Henry S. Thompson, Anne Anderson, Ellen Gurman Bard, Gwyneth Doherty-Sneddon, Alison Newlands, and Cathy Sotillo. 1993. The hcrc map task corpus:

Natural dialogue for speech recognition. In *Proceedings of the Workshop on Human Language Technology*, HLT '93, pages 25–30, USA. Association for Computational Linguistics.

Ashish Vaswani, Noam Shazeer, Niki Parmar, Jakob Uszkoreit, Llion Jones, Aidan N Gomez, Ł ukasz Kaiser, and Illia Polosukhin. 2017. Attention is all you need. In I. Guyon, U. V. Luxburg, S. Bengio, H. Wallach, R. Fergus, S. Vishwanathan, and R. Garnett, editors, *Advances in Neural Information Processing Systems 30*, pages 5998–6008. Curran Associates, Inc.

Wenhan Xiong, Xiaoxiao Guo, Mo Yu, Shiyu Chang, Bowen Zhou, and William Yang Wang. 2018. Scheduled policy optimization for natural language communication with intelligent agents. In *Proceedings of IJCAI*, pages 4503–4509.

Kelvin Xu, Jimmy Lei Ba, Ryan Kiros, Kyunghyun Cho, Aaron Courville, Ruslan Salakhutdinov, Richard S. Zemel, and Yoshua Bengio. 2015. Show, attend and tell: Neural image caption generation with visual attention. In *Proceedings of the 32nd International Conference on International Conference on Machine Learning - Volume 37*, ICML'15, page 2048â2057. JMLR.org.

Yanchao Yu, Arash Eshghi, and Oliver Lemon. 2017. Learning how to learn: An adaptive dialogue agent for incrementally learning visually grounded word meanings. In *Proceedings of the First Workshop on Language Grounding for Robotics*, pages 10–19, Vancouver, Canada. Association for Computational Linguistics.

Visually Grounded Follow-up Questions: a Dataset of Spatial Questions Which Require Dialogue History

Tianai Dong[1], Alberto Testoni[2], Luciana Benotti[3], Raffaella Bernardi[1,2]
[1] CIMeC, University of Trento, Italy [2] DISI, University of Trento, Italy
[3] Universidad Nacional de Córdoba, CONICET, Argentina
{tianai.dong|alberto.testoni|raffaella.bernardi}@unitn.it
luciana.benotti@unc.edu.ar

Abstract

In this paper, we define and evaluate a methodology for extracting history-dependent spatial questions from visual dialogues. We say that a question is history-dependent if it requires (parts of) its dialogue history to be interpreted. We argue that some kinds of visual questions define a context upon which a *follow-up spatial question* relies. We call the question that restricts the context: *trigger*, and we call the spatial question that requires the trigger question to be answered: *zoomer*. We automatically extract different trigger and zoomer pairs based on the visual property that the questions rely on (e.g. color, number). We manually annotate the automatically extracted trigger and zoomer pairs to verify which zoomers require their trigger. We implement a simple baseline architecture based on a SOTA multimodal encoder. Our results reveal that there is much room for improvement for answering history-dependent questions.

1 Introduction

The development of multimodal conversation agents is a long standing challenge (e.g. (Winograd, 1972)). In recent years, much has been achieved on the challenge of Visual Question Answering (VQA) (e.g. (Antol et al., 2015; Goyal et al., 2017).) The rapid advancements have brought researchers to further increase the difficulty of the task by proposing Visual Dialogue datasets (e.g. (Das et al., 2017; de Vries et al., 2017)) suitable to train multimodal dialogue systems. With this switch from VQA to Visual Dialogue, the challenge has increased in difficulty. First of all, while VQA involves only understanding the multimodal input (image and question), Visual Dialogues also require visual question generation and the acquisition of a dialogue strategy. Moreover, while VQA involves visual grounding of the question to be answered, Visual Dialogues require grounding the question against

Questioner	Oracle
Q1. Is it a fruit?	Yes
Q2. Is it in the foreground?	No
Q3. Are there two of them on the branch?	Yes
Q4. Is it the top one?	Yes

Figure 1: Through the dialogue the focus shifts from all the mandarins to just one. To answer Q4 (the `zoomer`), "top" needs to be interpreted relatively to the group of two mandarins identified by Q3 (the `trigger`).

both the visual and language contexts. As such this multi-folded challenge is rather ambitious. Our work focuses on identifying Follow-up Questions (FuQs) in Visual Dialogue. Namely, our goal is to construct a dataset of questions that we know require grounding both on the visual input and the dialogue history.

Work carried out on modeling the role of dialogue history in visual dialogue (Agarwal et al., 2020) has used the chit-chat dialogues of VisDial (Das et al., 2017) as a case study. However, it has been shown that in this dataset the role of grounding the question on the dialogue history is limited: models that take history into account do better, but the dataset contains a small percentage of questions that require dialogue history to be interpreted correctly. Based on these findings, Agarwal et al point out the need for data which captures dialogue history dependence. Our work is a contribution to this data collection challenge. We aim to identify FuQs which require (part of) the dialogue history to be interpreted.

Schlangen (2019) claims that goal-oriented settings will contain more dialogue phenomena. Following this claim, we run our analysis on GuessWhat?! (de Vries et al., 2017), a multimodal dataset in which the goal of the dialogues is to identify a referent in an image.

Proceedings of Second International Combined Workshop on Spatial Language Understanding andGrounded Communication for Robotics, pages 22–31
August 5–6, 2020.

In referential dialogues, the questions aim to collect information so as to narrow the set of potential candidates and univocally identify the referent among them. Interestingly, in referential multimodal game, this progressive refinement happens both through the language and visual contexts by incrementally zooming the joint attention to the conjectured referent. For instance in Figure 1, Q1 focuses on the full image and all objects are potential candidates. As the dialogue proceeds the attention is moved on the mandarins (Q1), then on those two mandarins in background (Q2) and finally on the mandarin on the top of the group of mandarins in the lower branch (Q3 and Q4). In this view, the data collection challenge launched in (Agarwal et al., 2020) can be rephrased by looking for a method to extract FuQs which require to zoom on a specific region of the image by narrowing the set of entities on which the dialogue focuses. We claim that FuQs requiring multimodal grounding can be extracted by identifying patterns of `trigger-zoomer` questions.

By manual inspection of the human dialogues, we have observed that often, after a positively answered question, the questioner tries to narrow down the choice by asking further details that discriminate the candidate. This happens in particular, when first a question (the `trigger`) identifies a group of objects that share some property and then the FuQ (the `zoomer`) focuses on one or more of the members of the identified referential set. In most cases, the `zoomer` requires the dialogue history to be answered. For instance, in Figure 1, the positively answered Q3 acts as a trigger which identifies the group of oranges under discussion and Q4 zooms on one of those. Notice that the question Q4 would be answered incorrectly if answered without considering Q3 and Q2 because the referent is not at the top of the picture. In this paper, we investigate the role of spatial questions in the identification of such patterns and focus on the evaluation of the Oracle player of the GuessWhat?! game. We show that the method we propose facilitates data collection of follow-up questions that need to be grounded on the visual and dialogue context to be answered correctly or at least with higher confidence. The dataset is publicly available at `https://github.com/tianaidong/2021SpLU-RoboNLP-VISPA` for future model developments and evaluations.

2 Related Work

Clark (1996) defines dialog *common ground* to be the commitments that the dialog partners have agreed upon during the dialog. An important part of the common ground is the *Question under Discussion (QuD)* (Ginzburg, 2012; De Kuthy et al., 2020). QuD is an analytic tool that has become popular among linguists and language philosophers as a way to characterize how a sentence fits in its context (Velleman and Beaver, 2016). The idea is that each sentence in discourse is interpreted with respect to a QuD. The QuD is defined by the dialog or discourse history. The linguistic form and the interpretation of an utterance, in turn, may depend on the QuD that provides the constraints that define the utterance's context. We reinterpret this theory to analyse referential visual dialogue: we take the QuD to be the objects conjectured to be the target. The interpretation of a question depends on its QuD.

Most of the work on the GuessWhat?! game has focused attention on the Questioner player; as a consequence, the issue of dialogue history needed by the Oracle has never been considered. Since the first baseline model (de Vries et al., 2017), the Oracle receives just the question without the previous turns. Furthermore, this baseline model is blind: it takes the question, the target's category and its location as inputs. This simple model has been widely used as the *Oracle* agent by all work on the *Questioner* (eg. (Strub et al., 2017; Shekhar et al., 2019; Pang and Wang, 2020).) Testoni et al. (2020) compared the LSTM baseline with a visually grounded LSTM (V-LSTM) and with an adaptation of LXMERT (Tan and Bansal, 2019). They show LXMERT based Oracle improves over the baseline achieving a new SOTA for the GuessWhat?! Oracle. Yet the model does not use dialog history as an input. We evaluate LSTM, V-LSTM and LXMERT against our dataset of context-dependent questions.

Agarwal et al. (2020) argues that although complex models that encode history for visual dialogs have been proposed (Yang et al., 2019), such work has not demonstrated that history matters for visual dialogs. Agarwal et al. propose and apply a new methodology for evaluating history dependence of questions in visual dialog. They show crowdsourcers a question with its image without the dialog history and ask the crowdsourcer "would you be able to answer this question by looking at the image only or you need more information from the previous conversation?”. However, it could happen that workers could be confident in answering the question just by looking at the image, but that they would give a different answer if the dialogue history is provided. This difference is crucial for studying context-dependent questions. In this paper, we proposed a new methodology for detecting history-dependent visual questions.

3 Dataset

We aim to identify Follow-up Questions (FuQs) that need the previous turn to be answered correctly or at least with higher confidence. We claim that FuQs which zoom in a specific region of the image to identify an object (or a set of objects) in it satisfy this request. This might hold in particular when the region contains more objects of the same category (e.g., more instances of mandarins, as in Figure 1) and the question refers to one (or more) member(s) of such a group. Moreover, we conjecture that most of such questions might also need to be visually grounded since the answer to it could change if the specific visual region they refer to is not properly identified and the question is mistakenly grounded over the full image. These challenging questions that zoom into a group are usually triggered by a question that refers to the whole group, the latter is identified by its location, the number or the color of its members. For instance, in Figure 1 the question that zooms on the target object of the game, “Is the top one?’, is triggered by the previous question that identifies the group itself by referring to the number of its members “Are there two of them on the branch?”. Interestingly, the zoomer question would be answered incorrectly without the previous turn since the target is at the top of the zoomed region and not on the top of the full image.

We focus on games in the test set in which there are more candidates of the same category of the target; we obtain 13,024 unique games containing 57,241 questions. We refer to these questions as the full test set. Shekhar et al. (2019) has classified GuessWhat?! questions into entity and attribute questions, the latter are subdivided into spatial, color, action, size, texture and shape. Testoni et al. (2020) further divided the spatial questions into group, absolute and relational questions. We build on these classifications to extract trigger and zoomer pairs. We see group and color questions as potential triggers for collecting history-dependent questions: for instance, group questions that contain explicit numbers indicate groups (e.g., "One of the three oranges?" refers to a group containing 3 members) and color questions might identify a group of objects which differ with respect to the color (e.g., "Is it blue?" may refer to a group of objects one of which is blue). For the zoomer questions, we consider group and absolute questions. Absolute questions are those spatial questions that contain an absolute location adjective (e.g., "Is it in the middle?" contains "middle"). Other types of questions, such as size ("Is it one of the big bottles?" which contains "big") and shape ("Is it kind of round?" which contains "round") could be used as triggers and zoomers as well. In this paper we do not use them because they are not frequent in the Guesswhat?! dataset and a preliminary analysis showed we would not extract sufficient trigger zoomer pairs through them.

Using the automatic annotation of Testoni et al. (2020), which is based on keyword matching, we extract group and absolute questions, 4342 and 11,743, respectively. Moreover, we extract the dialogues containing context-dependent group questions using the following patterns: a positively answered group question followed by another group question (Group-Group) and a positively answered color question followed by a group question (Color-Group); we obtained 364 and 145 pairs, respectively; and similarly for absolute questions obtaining 919 context-dependent absolute questions (530 from the Group-Absolute and 389 from the Color-Absolute patterns).

We randomly retrieve 200 samples for each subset[1] and manually checked them. We filtered out those pairs in which the zoomer question could be correctly interpreted without the dialogue history. We also removed samples that were noisy (the image was blurry or the target was too small, the question was not clear, etc). Each datapoint was annotated by two annotators (the four authors), and we maintained only those on which there was an agreement between the two annotators. After this filtering, we obtained in total 271 context-dependent questions manually checked: 164 group questions (103 group-group and 61 color-group) and 107 absolute questions, the latter are all from the group-absolute pattern.[2]

[1]For the Color-Group we took all the 145 datapoints.

[2]We are not considering questions extracted by color-absolute pattern in our evaluation, because the manual inspection of 200 samples randomly chosen from the automatically

We will refer to the set of visually grounded spatial questions that are context-dependent as VISPA. To gain a better understanding of the linguistic features of our dataset, we collect the statistics of question length, nouns and function words (*prepositions pronouns, determiners, conjunctions, auxiliaries*) for questions in each subset.[3] As we can see in Table 1 and Table 2, the context-dependent group and absolute questions do not show distinguishing surface features from the questions of the same type. Therefore they would have not be captured by using surface heuristics, such as searching for pronouns.

3.1 Examples

Figure 2 and Figure 3 report examples of context-dependent questions we have identified through our automatic process and further manual filtering. As we can see, when the previous turn is given, we can be much more confident in providing a correct answer. The previous turn is the question we have used to trigger the context-dependent FuQ, in one case its a group question ("is it between the two players in black?' "Yes"') and in the second case it is a color question ("one of the two gray ones?" "Yes"). The example on the upper part (group-group) is particularly interesting since the FuQ further specifies the previous turn, hence it should be properly integrated with it and interpreted as saying "Is it between the two players in black closest to the bat?". Only models that truly ground questions within the previous linguistic context can properly answer it. The latter example requires the Oracle to understand the group of objects the question refers to, the previous turn identifies this group through the color of its members.

Figure 3 provides an example of absolute questions in our manually filtered subset; the zoomer question would be answered negatively if the previous turn is not given, since "middle" would refer to the middle of the image. When the previous turn is given, "middle" should be instead interpreted as the middle of the 3 planes in front. This FuQ should be grounded on the linguistic and visual context to be properly answered.

4 Models

LSTM The first model we consider is the language-only baseline model proposed in (de Vries et al., 2017). This Oracle model receives as input the embeddings of the target object's category, its spatial coordinates, and the question to be answered encoded via an LSTM network. These three embeddings are concatenated and fed to a Multi-Layer Perceptron that gives the answer (Yes, No, or N/A).

V-LSTM We also consider a multimodal Oracle model. V-LSTM (Testoni et al., 2020) receives as input the embeddings of the target object's crop features, its spatial coordinates, the features of the image, and the question to be answered encoded via an LSTM network. All these embeddings are concatenated as in LSTM. The visual features are extracted with the frozen ResNet-152 network pre-trained on ImageNet (Russakovsky et al., 2015). Differently from LSTM, this model does not have access to the target object category.

LXMERT We additionally considered the Oracle model proposed in Testoni et al. (2020). This model is based on LXMERT (Learning Cross-Modality Encoder Representations from Transformers)(Tan and Bansal, 2019), a powerful multimodal transformer-based model. LXMERT represents an image by the set of position-aware object embeddings for the 36 most salient regions detected by a Faster R-CNN, and it processes the text input by position-aware randomly-initialized word embeddings. Both the visual and linguistic representations are processed by a specialized transformer encoder based on self-attention layers; their outputs are then processed by a cross-modality encoder that through a cross-attention mechanism generates representations of the single modality (language and visual output) enhanced with the other modality as well as their joint representation (cross-modality output). LXMERT uses the special tokens CLS and SEP. Testoni et al. (2020) fine-tuned the pre-trained version of LXMERT on the GuessWhat?! Oracle task by feeding the visual features and the spatial coordinates of the target object as the last region in the visual input. They took the representation corresponding to the special token CLS and fed it to a Multi-Layer Perceptron to obtain the answer to the input question. The authors show that this model outperforms the baseline model to a large extent.

extracted color-absolute questions provided too few cases of history-dependent questions.

[3] We utilize NLTK Python Package for the analysis

	Nr	Length	Nouns	Function W	Pronoun
All questions	57,241	4.89	1.23	3.08	0.75
Group Q	4342	7.27	1.51	4.31	0.61
Absolute Q	11,743	5.79	1.52	3.57	0.64
CD group Q	509	7.28	1.45	4.28	0.62
Group-Group	364	7.03	1.36	4.16	0.58
Color-Group	145	7.92	1.66	4.59	0.71
CD absolute Q	919	5.95	1.46	3.75	0.65
Group-Absolute	530	5.84	1.43	3.73	0.63
Color-Absolute	389	6.10	1.50	3.77	0.68

Table 1: Automatically extracted datapoints: Length: average question length; Nouns: average number of nouns per question; Function W: average number of function words per question; Pronouns: average number of pronouns per question

	Nr	Length	Nouns	Function W	Pronoun
CD Group Q					
Group-Group	103	6.89	1.26	4.10	0.66
Color-Group	61	7.50	1.49	4.55	0.72
CD Absolute Q					
Group-Absolute	107	5.59	1.31	3.76	0.57

Table 2: Manually filtered questions: Length: average question length; Nouns: average number of nouns per question; Function words: average number of function words per question; Pronouns: average number of pronouns per question

5 Experiments

We evaluated the models described above when receiving just the question or the question and the previous QA turn, we refer to the latter setting by marking the model names by -DH. We run each model three times (seed: 1, 50 and 100) and report their average together with the significance test results about the difference across runs. Table 3 and Table 4 report the model task accuracy on automatically extracted sets and manually filtered sets, respectively.

We claim the FuQs identified through the trigger-zoomer patterns need (at least) the previous turn to be answered properly or at least with higher confidence, this need should be even stronger for the manually filtered subsets.

As expected, LXMERT is the model that reaches the highest accuracy of the full test set (Table 3). Our results confirm what had been noticed by Testoni et al. (2020), namely that spatial questions are harder than average, and group questions are harder than absolute questions. This is reflected both by the baseline and the SOTA model: LSTM drops from 77.31 (All) to 70.45 (Absolute) to 67.11 (Group) and similarly does LXMERT – from 82.40 to 79.42 to 74.48. Even the accuracy of the best performing model, LXMERT-DH, further drops on the context-dependent questions reaching 74.43 and 71.12 for absolute and group questions, respectively.

When looking at the context-dependent questions, the standard-deviation among the accuracies reached by the three runs is rather high, hence in order to understand its effect on the comparison between models when receiving just the question and the question together with the previous turn, we have run a statistical significance paired t-test (following the suggestions in Dror et al. (2018).) The result shows that the difference between the two settings is never significant, except for LSTM/LSTM-DH on the absolute questions ($p\text{-value} < 0.05$). This shows that model performance is rather unstable and hence the selection of the binary answer is not properly grounded. This instability is not due to the size of the set: we have computed accuracy of the three runs of LXMERT on subsets of 500 and 100 randomly chosen questions and obtained a very low standard deviation.

Since the questions we have accurately selected are context-dependent, ideally a model should increase confidence in its answers when receiving the context (we simplify by giving just the the previ-

Questioner **Oracle**
Q10. Is it between the two players in black? Yes
Q11.The two players closest to the bat? Yes

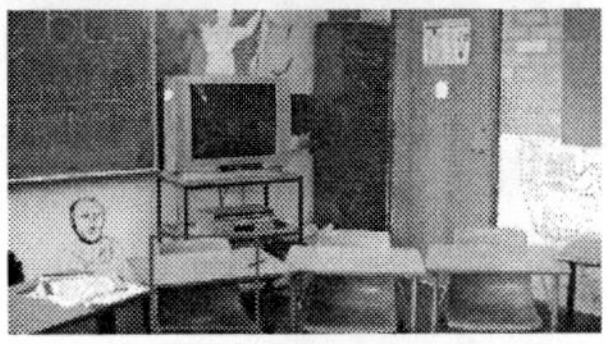

Questioner **Oracle**
Q4. one of the three gray ones? Yes
Q5. first one counting from left to right? Yes

Figure 2: Context-dependent group questions: group-group (up) color-group (down)

Questioner **Oracle**
Q2. One of the 3 planes in front? Yes
Q3. is the one in the middle? Yes

Figure 3: Context-dependent absolute questions: group-absolute

ous turn). To verify this hypothesis, we computed the confidence of LXMERT/LXMERT-DH (see Table 5) by using the average probability assigned to the answers in the manually filtered set. In our results, we find rather the opposite of our assumption: while both LXMERT and LXMERT-DH show relatively high confidence (>0.80) in providing correct answers, LXMERT-DH's confidences do not increase on LXMERT with the addition of the previous turn. On the positive side, we observe that for those cases where the model failed to provide the correct answer, LXMERT-DH is usually more uncertain than LXMERT about its own predictions. We consider this as a positive behavior of the model, since it suggests it is "aware" of what it does not know.

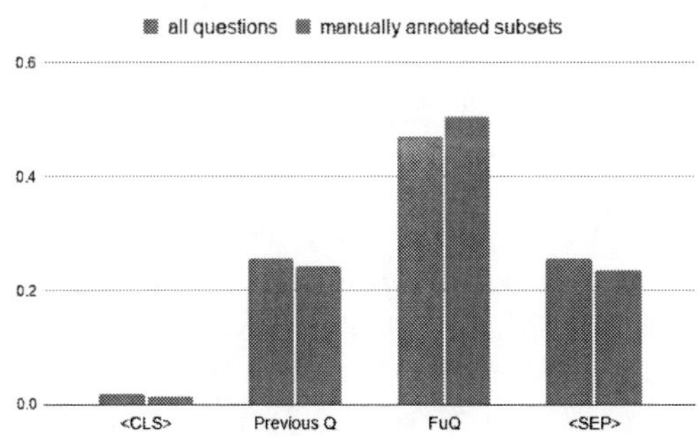

Figure 4: LXMERT-DH attention: all questions vs. context-dependent FuQs and the previous turn.

5.1 LXMERT attention

To understand the possible reasons that prevent the model from learning to exploit the dialogue history, we have analysed how LXMERT-DH puts attention to different parts of the input sequence through the computation of the cross-attention layers from language to vision (Figure 4). Ideally, context-dependent questions would require the model to put more attention to the trigger questions compared to questions that could be answered without the context. Model's attention on the previous turn in the manually selected subset should therefore be higher than in the full test set, if the model takes advantage of it as it should. However, this does not happen with LXMERT-DH: its attention on the previous turn does not change, and it actually slightly increases its attention on the zoomer question instead. This result confirms our claim that the current Oracle architecture fails in exploiting the trigger question while answering context-dependent questions and suggests that the model should be designed and trained to better attend the dynamically changing multimodal context.

	Controlled sets			Context Depedent	
	All	Absolute	Group	Absolute	Group
LSTM	77.31	70.45	67.11	60.57*	59.39
LSTM-DH	77.88	71.03	67.75	64.09*	59.06
V-LSTM	74.65	70.87	67.42	58.43*	62.15
V-LSTM-DH	73.82	70.13	65.12	63.33*	62.27
LXMERT	82.40	79.42	74.48	74.21	70.01
LXMERT-DH	82.79	80.19	74.49	74.43	71.12

Table 3: Models task accuracy when they receive vs. do not receive the previous turn. Context-Dependent Absolute questions are the only one for which statistically significant difference is found whe the DH is taken into accout (t-test among the runs of LSTM/LSTM-DH and V-LSTM/V-LSTM-DH, p-value < 0.05).

	Group-Group	Color-Group	Group-Absolute
LXMERT	65.84	71.58	76.95
LXMERT-DH	66.34	74.32	76.63

Table 4: Task accuracy on manually filtered sets of (271) Context-Dependent questions.

5.2 Qualitative Analysis

We have looked into the errors LXMERT does in three runs and compared them with those made by LXMERT-DH runs. Figure 5 illustrates the trigger-zoomer pairs in images that contain a color trigger question followed by a zoomer group question. We report three examples, in the first one all three runs of LXMERT-DH answers correctly while in two runs LXMERT does not; while in the others two examples both models fail in all runs.

The first example includes the spatial question *1 that is in the left?* that is answered incorrectly by LXMERT without history. Without history, we suspect it is answered with "Yes" since the target is indeed on the left of the image. However, if the previous trigger turn, *are there 2 black cars? Yes*, is considered, the objects that are relevant to answering the spatial question are the 2 black cars; withing this group, the target is not on the left but on the right. LXMERT with our simple history encoding is able to answer this spatial question correctly.

The second and third examples include spatial questions that our simple history encoding cannot capture. The second one (*is it the 1st one from right?*) is a spatial question that orders the objects in the inverse order. Usually objects are ordered from left to right but this question counts from right to left. The third example includes a group with four objects in the question *is it in the center row of 4 birds?*. We hypothesize that larger numbers are harder to interpret and answer correctly for LXMERT.

Some of the questions in our history-dependent dataset VISPA could be answered correctly by a human without reading the trigger question since they (being an Oracle) have access to the identity of the target and its attributes (such as category, color, etc). For instance, in the second game in Figure 5 the target is the red light on the right of the image (in the green box). A human Oracle can correctly answer the question *is it the 1st one from the right* assuming it considers only the red lights in the image, but without being sure the questioner is also making this assumption.. We also consider these questions to be history-dependent because they can be answered with more certainty considering the trigger question and its answer. We think that investigating whether history-dependent models become more certain of their correct answers (for the wrong reasons) is an interesting line for future research.

6 Discussion and Conclusions

Visual Dialogues are an interesting challenge because of the interplay between the language and visual modality. When focusing on answering visually grounded questions in dialogues, the main challenge they pose in addition to visual question answering is the need of grounding the question against the dialogue history. In our work we define and evaluate a methodology for extracting visually grounded history-dependent spatial questions from visual dialogues.

Our methodology does not capture all history dependencies in the dataset but it assures that

	Color-Group		Group-Group		Group-Absolute	
	Succeded	Failed	Succeded	Failed	Succeded	Failed
LXMERT	80.97	72.72	84.68	83.45	85.95	73.23
LXMERT-DH	81.37	63.95	80.54	71.37	82.30	65.51

Table 5: Confidence of models in answering FuQs in the manually filtered set. Succeeded: computed over the subsets in which the model provides the correct answer; Failed: computed over the subsets in which the model gives the wrong answer.

those pairs that are identified are indeed history-dependent.

The "trigger-zoomer" methodology we propose is evaluated here on the Guesswhat?! dataset. One possible question is how generic and applicable this model is in longer and open-world dialogues. We think that this method can be extended to longer dialogues by making the "trigger-zoomer" recursive. Moreover, it could be extended to datasets that not only contain questions, but also other forms of language. As far as a "trigger" affects a zoomer question by requiring the dynamic change of the multimodal attention to properly interpret it, the trigger can take any form. For instance, the trigger could be provided in the form of a caption referring to a specific region of an image. We believe that the "trigger-zoomer" methodology would be applicable to all open-word subdialogues that focus on reference resolution. Reference resolution is a frequent task in dialogue which takes up a large part of the turns in domains that are complex or need search. See for instance da Silva Rocha and Paraboni (2020).

We release both the automatically extracted question pairs as well as the subset of such questions which have been manually verified for context dependence. Some of these questions cannot be answered correctly without the previous trigger turn or at least confidence in answering them should be higher when the previous turns are provided. We evaluate the simple oracle models proposed so far in the literature and show that the architecture does not profit from the previous turn as it should. We pose the problem of interpreting follow-up questions as an open problem for the community.

References

Shubham Agarwal, Trung Bui, Joon-Young Lee, Ioannis Konstas, and Verena Rieser. 2020. History for visual dialog: Do we really need it? In *Proceedings of the 58th Annual Meeting of the Association for Computational Linguistics*, pages 8182–8197, Online. Association for Computational Linguistics.

Stanislaw Antol, Aishwarya Agrawal, Jiasen Lu, Margaret Mitchell, Dhruv Batra, C. Lawrence Zitnick, and Devi Parikh. 2015. VQA: Visual question answering. In *International Conference on Computer Vision (ICCV)*.

Herbert Clark. 1996. *Using Language*. Cambridge University Press, New York.

Abhishek Das, Satwik Kottur, Khushi Gupta, Avi Singh, Deshraj Yadav, José M.F. Moura, Devi Parikh, and Dhruv Batra. 2017. Visual Dialog. In *Proceedings of the IEEE Conference on Computer Vision and Pattern Recognition (CVPR)*.

Kordula De Kuthy, Madeeswaran Kannan, Haemanth Santhi Ponnusamy, and Detmar Meurers. 2020. Towards automatically generating questions under discussion to link information and discourse structure. In *Proceedings of the 28th International Conference on Computational Linguistics*, pages 5786–5798.

Rotem Dror, Gili Baumer, Segev Shlomov, and Roi Reichart. 2018. The hitchhiker's guide to testing statistical significance in natural language processing. In *Proceedings of the 56th Annual Meeting of the Association for Computational Linguistics (Volume 1: Long Papers)*, pages 1383–1392, Melbourne, Australia. Association for Computational Linguistics.

Jonathan Ginzburg. 2012. *The Interactive Stance*. Oxford Press.

Yash Goyal, Tejas Khot, Douglas Summers-Stay, Dhruv Batra, and Devi Parikh. 2017. Making the V in VQA matter: Elevating the role of image understanding in Visual Question Answering. In *Conference on Computer Vision and Pattern Recognition (CVPR)*.

Wei Pang and Xiaojie Wang. 2020. Visual dialogue state tracking for question generation. In *The Thirty-Fourth AAAI Conference on Artificial Intelligence, AAAI 2020, The Thirty-Second Innovative Applications of Artificial Intelligence Conference, IAAI 2020, The Tenth AAAI Symposium on Educational Advances in Artificial Intelligence, EAAI 2020, New York, NY, USA, February 7-12, 2020*, pages 11831–11838. AAAI Press.

Olga Russakovsky, Jia Deng, Hao Su, Jonathan Krause, Sanjeev Satheesh, Sean Ma, Zhiheng Huang, Andrej Karpathy, Aditya Khosla, Michael Bernstein, et al. 2015. Imagenet large scale visual recognition

Questioner	Oracle
1. is it a car?	no
2. is it white color?	no
3. is it red?	no
4. black?	yes
5. *are there 2 black cars?*	yes
6. *1 that is in the left?*	no
1. is it a car?	no
2. is it something on a building?	no
3. is it some light?	no
4. is it the street light?	no
5. can you see 5 or 6 of them on the right side?	no
6. *ok..so it is a light but may be the red lamps?*	yes
7. *is it the 1st one from right?*	yes
1. is it a bird?	yes
2. *is it white?*	yes
3. *is it in the center row of 4 birds?*	yes
4. Is it second from the front?	no
5. is it third from the front?	yes
6. from the top white birds it is in second?	yes
7. is it the top first?	yes

Figure 5: The two questions in italics in each dialogue correspond to pairs that start with a color question and continue with a group question. The first is an example in which LXMERT-DH answers correctly while LXMERT does not. The second and third ones illustrate kinds of spatial questions that are too challenging for our simple history encoding.

challenge. *International journal of computer vision*, 115(3):211–252.

David Schlangen. 2019. Grounded agreement games: Emphasizing conversational grounding in visual dialogue settings. *CoRR*, abs/1908.11279.

Ravi Shekhar, Aashish Venkatesh, Tim Baumgärtner, Elia Bruni, Barbara Plank, Raffaella Bernardi, and Raquel Fernández. 2019. Beyond task success: A closer look at jointly learning to see, ask, and GuessWhat. In *Proceedings of the 2019 Conference of the North American Chapter of the Association for Computational Linguistics: Human Language Technologies, Volume 1 (Long and Short Papers)*, pages 2578–2587.

Danillo da Silva Rocha and Ivandré Paraboni. 2020. Building referring expression corpora with and without feedback. *Lang. Resour. Evaluation*, 54(4):875–891.

Florian Strub, Harm De Vries, Jeremie Mary, Bilal Piot, Aaron Courville, and Olivier Pietquin. 2017. End-to-end optimization of goal-driven and visually grounded dialogue systems. In *Proceedings of international joint conference on artificial intelligenc (IJCAI)*.

Hao Tan and Mohit Bansal. 2019. LXMERT: Learning cross-modality encoder representations from transformers. In *Proceedings of the 2019 Conference on Empirical Methods in Natural Language Processing and the 9th International Joint Conference on Natural Language Processing (EMNLP-IJCNLP)*, pages 5103–5114.

Alberto Testoni, Claudio Greco, Tobias Bianchi, Mauricio Mazuecos, Agata Marcante, Luciana Benotti, and Raffaella Bernardi. 2020. They are not all alike: Answering different spatial questions requires different grounding strategies. In *Proceedings of the Third International Workshop on Spatial Language Understanding*, pages 29–38, Online. Association for Computational Linguistics.

Leah Velleman and David Beaver. 2016. Question-based models of information structure. In Caroline Féry and Shinichiro Ishihara, editors, *The Oxford Handbook of Information Structure*. Oxford University Press.

Harm de Vries, Florian Strub, Sarath Chandar, Olivier Pietquin, Hugo Larochelle, and Aaron C. Courville. 2017. GuessWhat?! Visual object discovery through multi-modal dialogue. In *2017 IEEE Con-*

ference on Computer Vision and Pattern Recognition, pages 5503–5512.

Terry Winograd. 1972. Understanding natural language. *Cognitive Psychology*, 3:1–191.

Tianhao Yang, Zheng-Jun Zha, and Hanwang Zhang. 2019. Making history matter: History-advantage sequence training for visual dialog. In *2019 IEEE/CVF International Conference on Computer Vision, ICCV 2019, Seoul, Korea (South), October 27 - November 2, 2019*, pages 2561–2569. IEEE.

Modeling Semantics and Pragmatics of Spatial Prepositions via Hierarchical Common-Sense Primitives

Georgiy Platonov Yifei Yang Haoyu Wu Jonathan Waxman

Marcus Hill Lenhart K. Schubert

Department of Computer Science, University of Rochester
gplatono@cs.rochester.edu
{yyang99, hwu36, jwaxman2, mhill24}@u.rochester.edu
schubert@cs.rochester.edu

Abstract

Understanding spatial expressions and using them appropriately is necessary for seamless and natural human-machine interaction. However, capturing the semantics and appropriate usage of spatial prepositions is notoriously difficult, because of their vagueness and polysemy. Although modern data-driven approaches are good at capturing statistical regularities in the usage, they usually require substantial sample sizes, often do not generalize well to unseen instances and, most importantly, their structure is essentially opaque to analysis, which makes diagnosing problems and understanding their reasoning process difficult. In this work, we discuss our attempt at modeling spatial senses of prepositions in English using a combination of rule-based and statistical learning approaches. Each preposition model is implemented as a tree where each node computes certain intuitive relations associated with the preposition, with the root computing the final value of the prepositional relation itself. The models operate on a set of artificial 3D "room world" environments, designed in Blender, taking the scene itself as an input. We also discuss our annotation framework used to collect human judgments employed in the model training. Both our factored models and black-box baseline models perform quite well, but the factored models will enable reasoned explanations of spatial relation judgements.

1 Introduction

Prepositions in general and spatial prepositions in particular form a notoriously difficult lexical class because of their inherent vagueness and polysemy. Pragmatics plays crucial role in determining both which prepositions are licensed for usage in a given situation and the range of configurations (i.e., locations of the arguments) of which the licensed preposition holds true. Spatial senses of prepositions are sensitive to miscellaneous factors such as shapes and salience of the argument objects, presence of meronymy (part-of) relations, typicality, etc. *On* provides a good example of such a semantically rich preposition. When we say that one object is on another one, we strongly imply the relation of physical support between them. But support relation comes in many forms and occurs in diverse physical configurations:

a) an apple on the table
b) a book on the shelf
c) a picture on the wall
d) a fly on the ceiling
e) a shirt on the person
f) a lamp on the post
g) a fish on a hook
h) a sail on a ship

Such variety makes capturing the meaning in a computational model difficult. Yet, locative expressions involving prepositions are pervasive in natural languages and, therefore, interpretation and understanding of their meaning is important for AI, especially in use cases involving grounded human-machine interactions. Another important requirement for modern AI systems is interpretability and explainability. While neural networks can efficiently learn complex statistical distributions from large datasets, they are predominantly opaque from the common-sense analysis perspective.

Our approach to computational models for spatial prepositions is based on the following considerations. To begin with, even though the range of senses of spatial relations together with the heavy dependence on pragmatic considerations make capturing their meaning with simple mathematical criteria difficult, it is still possible to account for many of the above aspects in a principled way. People's judgments about whether a particular relation holds

Proceedings of Second International Combined Workshop on Spatial Language Understanding andGrounded Communication for Robotics, pages 32–41
August 5–6, 2020.

in a given case can be quite variable; therefore it should suffice to provide models that estimate the probability that arbitrary judges would consider the relation to hold. This approach is aligned with a view of predicate vagueness as variability in applicability judgments (Kyburg, 2000; Lassiter and Goodman, 2017), enabling Bayesian interpretation. Next, since the usage of locative expressions is pragmatic, the ultimate success criterion in assessing models of prepositional predicates should also be pragmatic; i.e, in physical settings we often use such predicates to identify a referent (*the blue book in front of the laptop*) or to specify a goal (*put the laptop on the table*), so our models should allow a natural language system to interpret such usages as a human would.

Last, but not least, our approach facilitates explainability. Each relation is built from a combination of simpler relations, whose value can be retrieved and used to provide a justification for a particular judgement. For example, in order for one object to be next to another, they need to be close to each other and at about the same elevation. Thus, the latter criteria are included as factors in determining the value of the *next-to* relation, and their values could be used to generate meaningful explanations for any particular judgement made by the model.

In the following sections, we discuss related work, and then outline our modeling framework by examining the primitive concepts that are used as building blocks, and showing how these concepts come together in modeling a specific preposition. We then evaluate our approach in a "room world" domain, making use of Blender graphics software. We discuss two different sets of models, one purely neural network-based, implemented as a collection of multi-layer perceptrons, and another where models are implemented as trees, where each node computes a probabilistic rule. We describe our annotation framework for collecting human spatial judgments and evaluate our models. We summarize our contributions, and directions for future work, in the concluding section.

2 Related Work

In what follows, the first and second arguments of a preposition are referred to as *figure* and *ground*, respectively, when used in locative settings (Talmy, 1975).

The 3D approach to modeling spatial relations, as opposed to modeling based on 2D images, is informed by the cognitive science perspective. It is likely that people conceptualize their immediate surroundings as a 3D space defined by the three principal orientation axes of the body (Tversky et al., 1999). Moreover, 2D map-like space representations employed in navigation can be easily computed from a 3D "mental image" of the environment. It seems reasonable to assume that a potential embodied agent, such as robot, would also benefit from constructing such 3D "mental images" of its surroundings. Indoor scenarios for spatial modeling are particularly conducive to such approach (Bower and Morrow, 1990).

Developing computational models for spatial prepositions is a long-standing problem in the field of computational linguistics and NLP, and the attempts date back to the late 1960s. Early work followed mainly geometric intuitions, relying on the concepts of contiguity, surface, etc. (Cooper, 1968). A very good review of the semantic and pragmatic issues involved in spatial expressions is contained in Herskovits (1985). Herskovits' analysis identified a variety of important factors that influence correctness judgments in the application of spatial prepositions, illustrating these factors with many striking examples (e.g., the role of object types and typicality in contrasts such as *the house on the lake* vs. **the truck on the lake*, or the role of the figure/ground distinction and object size and type in contrasts like *The bicycle is near Mary's house* vs. *?Mary's house is near the bicycle*). Herskovits also proposed various abstract principles constraining the meaning and use of spatial prepositions. Our work borrows many of the elements of Herskovits' analysis, but is more narrowly focused on application to a particular setting (the room world), and is distinguished by our emphasis on developing computational models capable of actually evaluating the truth of prepositional relations in the chosen domain.

A number of methodologies rooted in application of topological notions to defining semantics of spatial prepositions arose aiming at spatial reasoning using abstract qualitative primitives to encode relations between objects (Cohn and Renz, 2008; Cohn, 1997). One example of such an approach is the Region Connection Calculus (RCC) and its modifications (Chen et al., 2015; Li and Ying, 2004). At the heart of RCC lies the notion of connectedness. Two nonempty regions are connected if and only if their topological closures have a nonempty intersection. Starting with this primitive, one may proceed to define more useful spatial relations such as part-of (x is a part of y if every object that is connected to x is also connected to y) and overlapping (x and y overlap if there is a z that is a part of both x and y). Continuing in the same fashion one can define several other topological notions and then use them to describe spatial configurations of objects. While mathematically appealing and facilitating rigorous inference, these qualitative methods are too strict and unable to capture the semantic richness of natural language descriptions of spatial configurations of objects, since they neglect aspects such as orientation, size, shape, and argument types.

Conceptually, the way we define the spatial relations in our model is similar to the *spatial template* approach, discussed in Logan and Sadler (1996). This approach is based on the idea of defining a region of acceptability around the reference object that captures the typical locations of the relatum for this relation and determining how well the actual relatum fits this region. Our work is also similar in spirit and goals to the work by Bigelow et al. (2015), which combined the imagistic space representations with spatial templates and applied it to a story understanding task. In their approach, the authors used explicit Blender graphics modeling of a scene to represent the objects in question and their relative configurations. In their model, each region of acceptability is a three dimensional rectangular region (more precisely, a prism with a rectangular base) representing the set of points for which the given spatial relation holds. For example if one has a pair of two objects, A and B, and wants to determine whether A is on top of B, A is checked to determine whether it is in the region of acceptability located directly above B. Probabilistic reasoning is supported by using values from 0 to 1 to represent the portion of the relatum that falls into a particular region of acceptability.

In recent years, attempts have been made to use statistical learning models, especially deep neural networks, to learn spatial relations. The work by Bisk et al. (2018) is concerned with learning to transduce verbal instructions, e.g., *"Move the McDonald's block so it's just to the right (not touching) the Twitter block"* into block displacements in a simulated environment. This system, unlike ours, relies on deep learning and does not use high-level cognitively-motivated spatial relation models. The CLEVR dataset (Johnson et al., 2017) and its modified versions, such as (Liu et al., 2019), lays out an explicit spatial question answering challenge that has inspired a flurry of visual reasoning works, e.g., (Kottur et al., 2019) and (Mao et al., 2019), which achieves near-perfect scores on the CLEVR questions. Common shortcomings of these approaches are reliance on synthetic data of limited variety (only a few simple geometric shapes are present), two-dimensional image-based model of the world, very limited ground-truth models of spatial relations (e.g., *left* means any amount laterally to the left, regardless of depth or intervening objects, etc.), and use of domain-specific procedural formalisms for linguistic semantics.

Other noteworthy recent examples of dataset-driven work are (Chang et al., 2014) and (Yu and Siskind, 2017). The former inverts the learning problem, in a sense; the task was not to learn how to describe object relationships, but rather to automatically generate a scene based on a textual description. The latter employed models of spatial relations to locate and identify similar objects in several video streams.

We should separately mention the spatial modelling studies by Malinowski and Fritz (2014) and, especially, Collell et al. (2017), which apply deep neural networks to learning spatial templates for triplets of form (relatum, relation, referent). The latter work does this in an implicit setting, that is, it uses relations that indirectly suggest certain spatial configurations, e.g., *(person, rides, horse)*. Their model is capable not only of learning a spatial template for specific arguments but also of generalizing that template to previously unseen objects; e.g., it can infer the template for *(person, rides, elephant)*. These approaches, however, rely on the analysis of 2D images rather than attempting to model relations in an explicitly represented 3D world.

Our approach can be seen as an attempt at quantitative implementations inspired by the criteria that have been discussed in psychologically and linguistically oriented studies (Garrod et al., 1999; Herskovits, 1985; Tyler and Evans, 2003). Studies of human judgements of spatial relations show that overly formal qualitative models with sharp boundaries generally cannot do justice to the usage of locative expressions in natural settings. We previously mentioned a study (Bigelow et al., 2015) that applied 3D graphics scene modeling to a story understanding task, allowing reasoning about the relative configuration and visibility of objects in the scene. Another example of an imagistic reasoning system was implemented as part of the planning system for the robot Ripley (Roy et al., 2004). Ripley used three-dimensional representation of its body, operator and workspace, reconstructed from two-dimensional view coming from Ripley's cameras.

Our work is very similar in spirit and execution to (Platonov and Schubert, 2018) and (Richard-Bollans et al., 2020b,a). All these studies model prepositions using specially designed 3D environments in Blender or Unity and employ similar sets of metrics to define the meaning of the prepositions. The studies by Platonov & Schubert differ from the present work in that the rules were less flexible (fewer parameters) and parameter values were hand-adjusted, while in our work we use gradient descent-based optimization to learn optimal values. The studies by Richard-Bollans *ete al.* relied on the prototype and exemplar approaches, using learning from data to estimate the prototype parameters or the exemplar configuration. Our work is, by contrast, rule-based (although one might argue that the parameters in our rules implicitly encode prototype properties). None of the prior studies explore generation of justifications for the spatial judgements.

3 Task Description

We explore spatial prepositions as applied to the so-called "room worlds" - 3D scenes depicting room interiors filled with common everyday items such as furniture, appliances, food items, etc.

The objects in the scene are designed in a particular way, so that their meronomy corresponds to that of the real objects. That is, the mesh consists of parts that are usually distinguished by people (e.g., for a chair, its seat, legs, back, ets., are separate objects that can be accessed by our system). This is useful for part-based inferences, e.g., a book is on a bookshelf when it is on one of the shelves. The objects are also annotated with other additional tags such as frontal vectors that indicate where the "front" of an object is, object type, etc. We have designed 52 scenes containing about 10-30 objects admissible for annotation as figure objects. Since our annotation task involves describing the location of a figure object in relation to other objects (grounds), objects that form the environment (walls, ceiling, floor) are not admissible as figures (however, they can be used as grounds as in *the poster is on the north wall*).

This serves as a realistic domain for evaluating spatial relations. We designed the annotation task so as to achieve a balance between obtaining a significant number of annotations and collecting some information about human preferences. In each annotation instance the annotator is presented with a screenshot of a room world scene and is asked to describe the location of a highlighted figure object. First, the annotator is to pick a single best-fitting preposition and a corresponding ground object. After that, they are to indicate all other relations that they believe to hold between the figure and the ground (if the most appropriate relation chosen was *between*, they are to indicate which relations hold between the figure and the first ground object). They are then asked to repeat the same procedure for up to two more times. The reason for such an approach is that while the second part of the annotation (choosing all the relations holding between the given and the selected objects) produces coverage of pairwise relations, many such judgements feel forced and unnatural to human annotators (during earlier explorations it was noted that vagueness of locative expressions leads to annotators overthinking when making judgements). The laws of conversational implicature predict that, in everyday usage, various locatives will not occur with uniform frequencies. When several possible prepositions are applicable, people tend to choose those prepositions that disambiguate better or imply stronger relations, e.g., *on* is preferable to *touching* or *near* even though these relations often co-occur. Hence, the first part of the annotation process allows annotators to freely choose the most natural or "obvious" options.

At the moment, because of the scarcity of data (see Table 2 for the number of collected annotations), we don't distinguish between the two annotation types when training and testing our models. In principle, one can assign different weights to different annotations to skew the model towards relying on the best-choice annotations more.

4 Model Details

We have developed two kinds of models. The first one is a series of simple multi-layer perceptrons (one per each relation), and the second is our main rule-based model, which is implemented as a net-

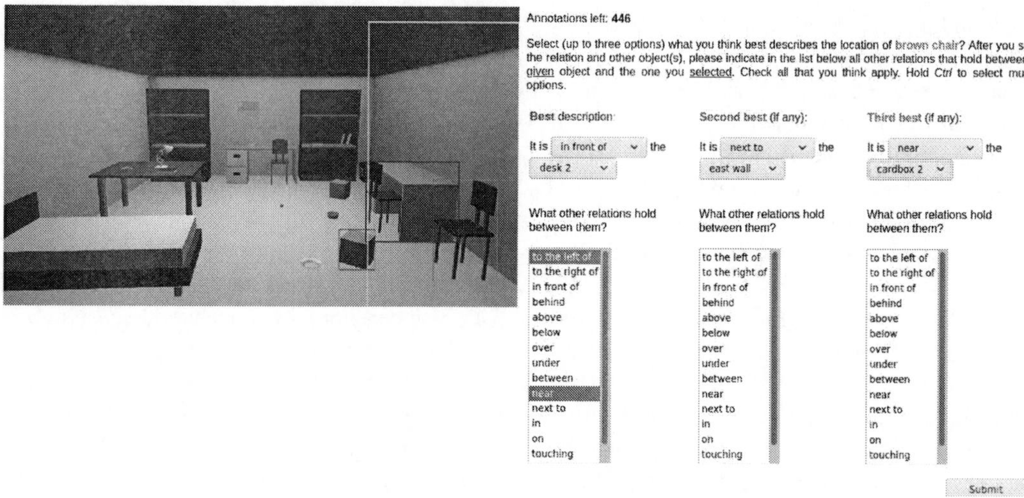

Figure 1: An example of a room world scene and the accompanying annotation controls. Best viewed in color.

work (more precisely, an arborescence) of nodes that compute meaningful hand-crafted relations used for determining the values of the prepositions. Each node realizes one or more differentiable operations which allows us to train the model using standard gradient descent-based optimization. The main reasons for developing the pure NN-based solution are to provide the baseline performance metric against which we compare our main models. Each model is essentially a binary classifier used to predict the likelihood that a particular relation holds between given objects.

4.1 Neural Baseline

Our baseline model consists of a number of independent binary classifiers (one for each spatial relation) and employs a 2-hidden layer architecture for each network. The baseline models take figure and ground objects' centroids, bounding boxes, and frontal vectors as input features. For each relation we iteratively tested different hidden layer structures in the 15-36 units range and selected one that performs the best (on average, across 5 randomized re-runs). We chose SELU activations (Klambauer et al., 2017) in the hidden layers and the logistic sigmoid function as an output non-linearity, which was the best combination based on our empirical exploration. We used binary cross-entropy as the standard binary classification loss. The model was trained using the PyTorch stochastic gradient descent optimizer with learning rate $\eta = 0.003$ and momentum $\alpha = 0.9$. We experimented with different regularization terms, but didn't notice any consistent performance gains (probably due to the small size of our networks and dataset). Main reason for the simplicity of the neural baseline is the small size of the dataset of annotations (under 7000 in total).

4.2 Rule-Based Model

We rely on a soft rule-based approach and imagistic scene representation for computing spatial relations. Each spatial preposition is implemented as a binary or ternary probabilistic predicate computed hierarchically as a combination of more primitive relations that we call *factors*. These factors encode typical more basic relations that affect whether a particular spatial preposition holds. They are usually either different senses of the same preposition or they co-occur with the preposition in most/all configurations that license the usage of that preposition. The set of factors ranges from those computing geometric properties (e.g., locations, sizes, and distances) to ones computing non-geometric, or functional ones (e.g., physical properties of the relata, such as part structure, or the location of the "front" of an object). There are several combinatory rules that determine how the factors are combined to produce a composite value. Typically, the factor values are linearly combined, multiplied together, or the maximum among them is taken, depending on the relation. For example, when one object is "on" another, it is often higher than the second object, and typically supported by it. The factors that we compute represent such primitive relations that often accompany higher-level relations of "on-ness", "above-ness", etc. A list of example factors is presented in the Table. 1 below.

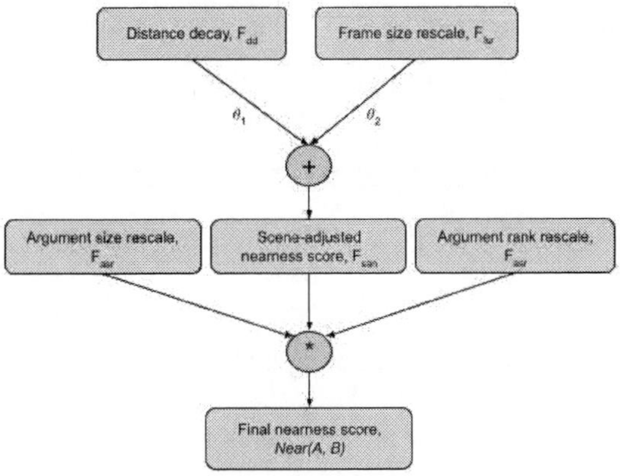

Figure 2: Structure of the factor network for *near*.

The factor tree for each relation is different, however, the general underlying principles can be understood by considering an example. One such example factor network is presented in Fig. 2. When computing *Near*(A, B), we start by computing the absolute distance, $d(A, B)$, between A and B. How this distance is computed depends on the geometry of the arguments. In the default case, assuming that both objects are roughly compact, $d(A, B)$ is simply the Euclidean distance between the centroids of A and B since, in this case, the centroid is a good approximation of the "general location" of an object. On the other hand, if, say, A or B is planar (extended in any two dimensions compared to the third, e.g., a wall, a book, a TV, etc.), linear (extended in one dimension, e.g., a pen), or generally concave (e.g., a table), then $d(A, B)$ is the minimum between the centroid distance and the distance between two closest points of A and B. We then compute scaled distance $d_{sc}(A, B)$ by dividing the absolute distance by the sum of the argument sizes, which are approximated by the radius of the circumscribed sphere. Intuitively, scaled distance provides a "size invariant" measure of the closeness of the two objects. Its value should be close to 1 when the objects are adjacent to each other, regardless of the their sizes. Next, we compute the *distance decay factor*, F_{dd}, as

$$F_{dd}(A, B) = \sigma(\theta_{dd} d_{sc}(A, B)),$$

where σ is logistic sigmoid and θ_{dd} is a learned parameter. The value of this factor gives a context-independent measure of nearness, which is then subsequently modified by a rescaling that takes into account context information. We compute the *scene-adjusted nearness*, F_{san}, as a linear combination

$$F_{san}(A, B) = \theta_1 F_{dd}(A, B) + \theta_2 F_{fsr}(A, B),$$

where $\theta_1, \theta_2 \geq 0, \theta_1 + \theta_2 = 1$, and

$$F_{fsr} = 1 - \frac{d(A, B)}{frame\_size}$$

is the *frame-size rescale factor*. The latter gives an estimate of nearness by considering the absolute distance between the objects relative to the size of the *frame*, i.e., psychologically salient part of the world. Currently, frame size is taken to be the size of the entire scene. However, in principle this can be extended to be chosen depending on argument locations, e.g., if two small objects are on top of a table, we can make the frame be the area of the table top. The final nearness score is computed as

$$Near(A, B) = F_{san}(A, B) F_{asr}(A, B) F_{arr}(A, B).$$

Here, F_{asr} is the *argument size-rescaling factor*,

$$F_{asr}(A, B) = 0.9 + 0.1 \cdot \sigma(\theta_{asr}(B.size - A.size)),$$

if $A.size > B.size$, and $F_{asr}(A, B) = 1$ otherwise. This factor encodes the intuition that, when using *near* to locate objects, the ground object is typically chosen to be bigger and fixed. Compare *?the house is near the car* vs. *the car is near the house*. Thus, when the figure is bigger than the ground we reduce the nearness score a bit, so that $Near(Bookshelf, Banana)$ returns a lower value than $Near(Banana, Bookshelf)$ (other things being equal). However, as should be clear from the formula for F_{asr}, we only allow the size difference adjustment to vary in the interval $[0.9, 1.0]$. In this way, the system would prefer to use the correct order of the arguments when making a nearness judgement on its own, while still recognizing that the relation might hold for the reverse order of the arguments.

The F_{arr} is the *argument ranking rescaling factor*. This factor lowers the nearness score if there are other objects that have a higher value of F_{san}. That is, it lowers the score in proportion to how far the current figure object is from being the best candidate figure object for a given selection of the ground and the relation. It is computed as

$$F_{arr} = e^{-\theta_{arr}(rank-1)},$$

Factor	Description
$to\_the\_right\_of\_deictic(a, b, o)$	Represents the deictic (here - viewer-specific) sense of the *to the right of* with respect to the observer o
$in\_front\_of\_intrinsic(a, b)$	Represents the intrinsic (object-centered) sense of *in front of*
$frame\_size\_rescale(a, b)$	Relative distance between a and b based on the size of the current perceptual frame
$supporting(a, b)$	Direct support relation, i.e., whether a supports b
$indirectly\_supporting(a, b)$	Indirect support relation, i.e., whether a supports some c which, in turn, supports b
$touching(a, b)$	Whether a and b are in contact with each other
$in\_direction(a, b, v)$	Computes whether b is in the general direction defined by a vector v with respect to a
$higher\_than\_centroidwise(a, b)$	Determines whether a is higher than b in terms of their centroid locations
$at\_same\_height(a, b)$	Computes whether a and b are roughly at the same elevation (in terms of centroids or their base level)

Table 1: Some of the factors used in computing spatial relations. In our system, we use the term observer to refer to the properties of the viewer, i.e., viewer location and gaze direction.

where $rank$ is the number of other objects C such that $F_{san}(C, B) > F_{san}(A, B)$.

Regarding sense ambiguity, different relations can be evaluated with respect to different coordinate frames. For example, for several projective relations, e.g., *to the right of*, we consider three cases, deictic, extrinsic and intrinsic. The so-called *deictic to the right of* is computed based on viewer's perspective. Here, one object is considered to be to the right of another, if its projection onto the viewer's visual plane is to the right of that of the latter. The *extrinsic to the right of* is based on the global coordinate system imposed by the world, i.e., front-right sides of the room. Finally, the *intrinsic to the right of* is determined based on the intrinsic coordinate system of the ground object, i.e., A is intrinsically to the right of B if it is on the right side of B. Note that not all objects have intrinsic orientations, and in these cases this sense of the relation is assigned 0. These different senses are evaluated based on the known observer properties (location and gaze direction), global orientation vectors of the world (fixed and always known), and frontal vector of an object (when applicable, i.e., the object has inherent orientation), respectively. When dealing with multiple senses, the model selects the one with the maximal value as an output.

The rule-based models are implemented as custom computational graphs using the PyTorch framework. We use binary cross-entropy loss and Adam as an optimizer, with the learning rate $\eta = 0.01$ and L2 regularization coefficient 0.1. The models are trained using back-propagation of error. Each object (3D mesh) in the scene is encapsulated in a separate Python object. It should be noted that we use these Python objects as input features, and not the numerical vectors as is common in the ML work.

5 Evaluation and Discussion

Evaluation data for both types of models are presented in Table. 2. Overall, both models performed reasonably well, apart from the cases such as *in front of, behind* and *touching* where the rule-based model performed better thanks to additional available information. The results clearly show that it is possible to produce reasonable judgments for most spatial relations even with purely geometric information. However, our main goal was to demonstrate that even when they fall short, our rule-based models still compare reasonably well with pure neural network-based approaches, with the added benefit of being interpretable thanks to their formulation in terms of meaningful decision criteria that correspond to human intuitions about spatial relations. Another important aim of our exploration was to evaluate whether the factors we selected are appropriate and sufficient for modeling the semantics of the locative senses of prepositions. While it is difficult to extrapolate our performance results to novel settings, we believe that our room worlds are representative of a significant subset of everyday settings where locative expressions are apt to be used. The annotation process is still ongoing and we are working on an additional set of scenes depicting outdoor environments. As such, the numbers in the table are subject to change, as the breadth of configurations covered and annotation data is increased. Scale differences between the two domains might affect the boundaries of applicability of the prepositions as well as their relative psychological preference ordering. Whenever possible we rely on approximations to the real 3D meshes of objects, using centroids and bounding boxes. This allows us to focus on the most salient features of objects' shapes and maintain relatively high performance. The system generates responses in real time which is relevant to the responsive-

		Pure NN model				Rule-based model			
relation	total instances	accuracy	precision	recall	F1	accuracy	precision	recall	F1
to the right of	214	0.94	1.00	0.89	0.94	0.94	0.97	0.92	0.94
to the left of	152	0.89	0.85	1.00	0.92	0.95	1.00	0.90	0.95
in front of	127	0.73	0.66	0.90	0.76	0.85	0.81	0.93	0.87
behind	97	0.76	0.68	0.91	0.78	0.86	0.80	0.91	0.85
above	74	1.00	1.00	1.00	1.00	0.90	1.00	0.85	0.92
below	86	0.82	0.92	0.80	0.85	0.87	0.97	0.78	0.87
between	220	0.96	1.00	0.93	0.96	0.95	1.00	0.87	0.93
next to	331	0.97	0.97	1.00	0.98	0.95	0.94	1.00	0.97
touching	82	0.76	0.74	0.83	0.78	0.99	1.00	0.97	0.98
near	296	0.90	0.91	0.95	0.93	0.93	0.95	0.93	0.94
on	346	0.8	0.81	0.89	0.85	0.89	0.94	0.88	0.91

Table 2: Performance statistics for the rule-based (RB) and pure MLP (NN) models. We excluded the data for *under, over* and *in*, as the number of collected annotations was insufficient. The total instances column refers to the test set instances, which constitute between 20% and 30% of all collected annotations, depending on the relation.

ness during a dialogue with the user (see the next subsection).

5.1 On Explainability

The main reason for using the rule-based approach is its interpretability. Specifically, our tree-of-factors implementation of spatial models allows backwards-generated justification of the final judgment. Each factor represents some higher-level semantic concept which can be readily translated into natural language. The tree of factors computed during the forward computation phase is preserved and is traversed in the backward direction starting from the root (representing the final output, i.e., the result of the evaluation of the preposition model). The mechanism for selecting the relevant factors for each node of the tree is as follows. If the combination rule for the current node (the way the factors of its immediate children are combined) is a product, then if the node value ≥ 0.5, return all the child nodes; otherwise, return the child node with the smallest value. If the combination rule for the children is a weighted linear combination of factor values, then if the current node value is ≥ 0.5, return the highest contributing factor node or nodes (total contribution includes their value and weight); otherwise, return the value of the node with the largest weight. Finally, if the combination rule is the max operation, then if the current node value is ≥ 0.5, return the child node with maximum value; otherwise return all the child nodes. One exception is the *touching* relation, for which the explanation procedure returns a particular part of the ground object as a justification (if the relation holds, that is). For example, the relation *Touching(Green Book, Bookshelf)* holds because the relation *Touching(Green Book, shelf 2)* holds, where *shelf 2* is part of the *Bookshelf*. In this case, relation between parts is considered a primitive, i.e., non-decomposable into more primitive relations, and so the justification process ends there. We are currently working on incorporating our models into an existing dialogue system that, given the returned factor(s), will generate an output in English. The interpretability of our models is to be evaluated in a dialogue-based setting.

As an example of the operation of the explanation procedure, consider simplified factor network for *to the right of* in Fig. 3.

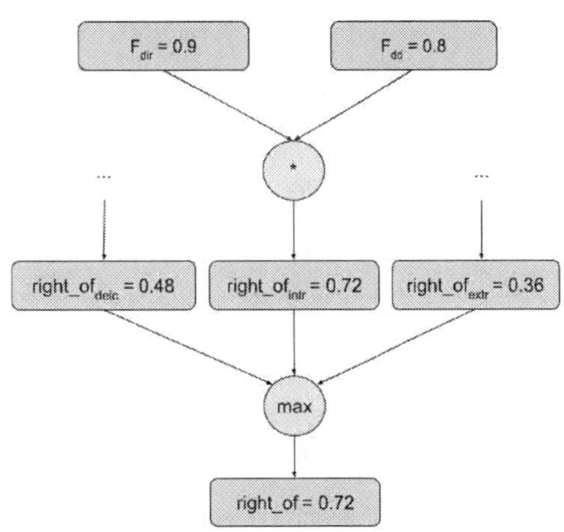

Figure 3: An example of an explanation procedure.

The numbers in the nodes are the respective values of the factors that the node computes. Assume that the system is being asked whether A is to the right of B. Assume further that the final output value is *right of* = *0.72*, which corresponds to "yes". Now, if the user inquires why the system arrived at that conclusion, the following process unfolds. The node for the final score for *to the right of* takes the maximum over three values: deictic $right\_of_{deic}$, intrinsic $right\_of_{intr}$ and extrinsic $right\_of_{extr}$. Since the maximum is taken, one of those nodes must be equal to the final value. Hence, the explanatory routine returns the corresponding node and its value ($right\_of_{intr}$, 0.72). The corresponding interpretation will be (when bridging with the dialogue system is completed) something like "A is to the right of B because A is located on the right side of B". If asked further as to why the intrinsic relation holds, the system will analyze the intrinsic score's contributing factors, namely F_{dir} (directional factor that defines the "right-side" region for an object) and F_{dd} (distance decay, measuring how far apart the objects are). Since the combination rule used is multiplication and the value of the current node (intrinsic right) is 0.72 (i.e., relation holds), it follows that both factors must hold as well. The system will return the list of the nodes and their values, i.e., [$(F_{dir}, 0.9)$, $(F_{dd}, 0.8)$] as a result. The straightforward interpretation of the latter would be "A is on the right side of B, because it is located in the general rightward direction w.r.t. to B and it is close enough". This process can continue until leaf nodes are reached, which do not admit further decomposition and are treated as primitives. Alternatively, assume that the value F_{dd} is only 0.4 (A is too far from B). This low value will propagate downstream and affect the intrinsic $right\_of_{intr}$ and the final *right of* score. In this case, the system will supply a negative answer to the original question. When asked why A is not to the right of B, it will return the list of all senses [($right\_of_{deic}$, 0.48), ...] which has a straightforward interpretation of "A is not to the right of B because none of the senses apply". If queried why, say, the intrinsic sense does not apply, the system returns the lowest-value node contributing to the intrinsic sense node, i.e., [$(F_{dd}, 0.4)$], which translates into "A is too far from B to be on its right side".

Note the contrast with standard approaches to explainability in deep neural networks (e.g., modular neural networks), where the model can usually only answer "what" questions about its decisions (i.e., we know what kind of thing a module computes), but not the "why" or "how" questions about the reasons a given module arrived at a particular output.

6 Conclusion

We considered the problem of designing intuitive computational models of spatial prepositions that combine geometrical information as well as some pieces of commonsense knowledge and contextual information about the arguments. Our main aim was to develop spatial semantic models that rely on psychologically plausible criteria and facilitate justification of spatial judgements produced by the models, and to compare such an approach against a more mainstream black box statistical learning architecture acting as a baseline. We believe that combining the power of data-driven methods and interpretable, algorithmic models is the way forward in AI in general and, in particular, is necessary in order to incorporate context and background knowledge information needed to model spatial expressions properly. This work is one step in that direction.

References

Eric Bigelow, Daniel Scarafoni, Lenhart Schubert, and Alex Wilson. 2015. On the need for imagistic modeling in story understanding. *Biologically Inspired Cognitive Architectures*, 11:22–28.

Yonatan Bisk, Kevin J Shih, Yejin Choi, and Daniel Marcu. 2018. Learning interpretable spatial operations in a rich 3d blocks world. In *Thirty-Second AAAI Conference on Artificial Intelligence*.

Gordon H Bower and Daniel G Morrow. 1990. Mental models in narrative comprehension. *Science*, 247(4938):44–48.

Angel Chang, Manolis Savva, and Christopher D Manning. 2014. Learning spatial knowledge for text to 3d scene generation. In *Proceedings of the 2014 Conference on Empirical Methods in Natural Language Processing (EMNLP)*, pages 2028–2038.

Juan Chen, Anthony G Cohn, Dayou Liu, Shengsheng Wang, Jihong Ouyang, and Qiangyuan Yu. 2015. A survey of qualitative spatial representations. *The Knowledge Engineering Review*, 30(01):106–136.

Anthony G Cohn. 1997. Qualitative spatial representation and reasoning techniques. In *KI-97: Advances in Artificial Intelligence*, pages 1–30. Springer.

Anthony G Cohn and Jochen Renz. 2008. Qualitative spatial representation and reasoning. *Foundations of Artificial Intelligence*, page 551.

Guillem Collell, Luc Van Gool, and Marie-Francine Moens. 2017. Acquiring common sense spatial knowledge through implicit spatial templates. *arXiv preprint arXiv:1711.06821*.

Gloria S Cooper. 1968. A semantic analysis of english locative prepositions. Technical report, BOLT BERANEK AND NEWMAN INC CAMBRIDGE MA.

Simon Garrod, Gillian Ferrier, and Siobhan Campbell. 1999. In and on: investigating the functional geometry of spatial prepositions. *Cognition*, 72(2):167–189.

Annette Herskovits. 1985. Semantics and pragmatics of locative expressions. *Cognitive Science*, 9(3):341–378.

Justin Johnson, Bharath Hariharan, Laurens van der Maaten, Li Fei-Fei, C Lawrence Zitnick, and Ross Girshick. 2017. Clevr: A diagnostic dataset for compositional language and elementary visual reasoning. In *Proceedings of the IEEE Conference on Computer Vision and Pattern Recognition*, pages 2901–2910.

Günter Klambauer, Thomas Unterthiner, Andreas Mayr, and Sepp Hochreiter. 2017. Self-normalizing neural networks. *arXiv preprint arXiv:1706.02515*.

Satwik Kottur, José MF Moura, Devi Parikh, Dhruv Batra, and Marcus Rohrbach. 2019. Clevr-dialog: A diagnostic dataset for multi-round reasoning in visual dialog. *arXiv preprint arXiv:1903.03166*.

Alice Kyburg. 2000. When vague sentences inform: a model of assertability. *Synthese*, 124:175–191.

Daniel Lassiter and Noah D. Goodman. 2017. Adjectival vagueness in a Bayesian model of interpretation. *Synthese*, 194:3801–3836.

Sanjiang Li and Mingsheng Ying. 2004. Generalized region connection calculus. *Artificial Intelligence*, 160(1):1–34.

Runtao Liu, Chenxi Liu, Yutong Bai, and Alan L Yuille. 2019. Clevr-ref+: Diagnosing visual reasoning with referring expressions. In *Proceedings of the IEEE Conference on Computer Vision and Pattern Recognition*, pages 4185–4194.

Gordon D Logan and Daniel D Sadler. 1996. A computational analysis of the apprehension of spatial relations. *Language and space*.

Mateusz Malinowski and Mario Fritz. 2014. A pooling approach to modelling spatial relations for image retrieval and annotation. *arXiv preprint arXiv:1411.5190*.

Jiayuan Mao, Chuang Gan, Pushmeet Kohli, Joshua B Tenenbaum, and Jiajun Wu. 2019. The neuro-symbolic concept learner: Interpreting scenes, words, and sentences from natural supervision. *arXiv preprint arXiv:1904.12584*.

Georgiy Platonov and Lenhart Schubert. 2018. Computational models for spatial prepositions. In *Proceedings of the First International Workshop on Spatial Language Understanding*, pages 21–30.

Adam Richard-Bollans, Lucía Gómez Álvarez, and Anthony G Cohn. 2020a. Modelling the polysemy of spatial prepositions in referring expressions. In *Proceedings of the International Conference on Principles of Knowledge Representation and Reasoning*, volume 17, pages 703–712.

Adam Richard-Bollans, Brandon Bennett, and A Cohn. 2020b. Automatic generation of typicality measures for spatial language in grounded settings. In *European Conference on Artificial Intelligence*. Leeds.

Deb Roy, Kai-Yuh Hsiao, and Nikolaos Mavridis. 2004. Mental imagery for a conversational robot. *Systems, Man, and Cybernetics, Part B: Cybernetics, IEEE Transactions on*, 34(3):1374–1383.

Leonard Talmy. 1975. Figure and ground in complex sentences. In *Annual Meeting of the Berkeley Linguistics Society*, volume 1, pages 419–430.

Barbara Tversky, Julie Bauer Morrison, Nancy Franklin, and David J Bryant. 1999. Three spaces of spatial cognition. *The Professional Geographer*, 51(4):516–524.

Andrea Tyler and Vyvyan Evans. 2003. *The semantics of English prepositions: Spatial scenes, embodied meaning, and cognition*. Cambridge University Press.

Haonan Yu and Jeffrey Mark Siskind. 2017. Sentence directed video object codiscovery. *International Journal of Computer Vision*, 124(3):312–334.

Towards Navigation by Reasoning over Spatial Configurations

Yue Zhang
Michigan State University
zhan1624@msu.edu

Quan Guo
Michigan State University
guoquan@msu.edu

Parisa Kordjamshidi
Michigan State University
kordjams@msu.edu

Abstract

We deal with the navigation problem where the agent follows natural language instructions while observing the environment. Focusing on language understanding, we show the importance of spatial semantics in grounding navigation instructions into visual perceptions. We propose a neural agent that uses the elements of spatial configurations and investigate their influence on the navigation agent's reasoning ability. Moreover, we model the sequential execution order and align visual objects with spatial configurations in the instruction. Our neural agent improves strong baselines on the seen environments and shows competitive performance on the unseen environments. Additionally, the experimental results demonstrate that explicit modeling of spatial semantic elements in the instructions can improve the grounding and spatial reasoning of the model.

1 Introduction

The ability to understand and follow natural language instructions is critical for intelligent agents to interact with humans and the physical world. One of the recently designed tasks in this direction is Vision-and-Language Navigation (VLN) (Anderson et al., 2018), which requires an agent to carry out a sequence of actions in a photo-realistic simulated environment in response to a sequence of natural language instructions. To accomplish this task, the agent should have three abilities: understanding linguistic semantics, perceiving the visual environment, and reasoning over both modalities (Zhu et al., 2020; Wang et al., 2019). While understanding vision and language are difficult problems by themselves, learning the connection between them without direct supervision makes this task even more challenging (Hong et al., 2020).

To address this challenge, some neural agents establish the connection using attention mechanism to relate the tokens from a given instruction to the images in a panoramic photo (Anderson et al., 2018; Fried et al., 2018; Ma et al., 2019; Yu et al., 2018). Surprisingly, although those models can improve the performance, Hu et al. (2019) found they ignore the visual information. There is no clear evidence that the agent can correspond the components of the visual environment to the instructions (Hong et al., 2020). Based on these results, recent research started to improve the agent's reasoning ability by explicitly considering the structure of language and image. From the language side, Hong et al. (2020) annotated fine-grained sub-instructions and their corresponding trajectories and used the co-grounded features of a part of instruction and the image to predict the next action. From the image side, Hu et al. (2019) induced a high-level object-based visual representation to ground the language into the visual context.

In the same direction, we propose a neural agent, namely *Spatial-Configuration-Based-Navigation (SpC-NAV)*, and consider the structure of both modalities, that is, spatial semantics of the instructions and the objects in the images. We use the notion of *Spatial Configuration* (Dan et al., 2020) to model the instructions and design a state attention to ensure the execution order of spatial configurations. Then, we utilize the spatial semantics elements, namely *motion indicator*, *spatial indicator* and *landmark* in spatial configuration to establish the connection with the visual environment. Specifically, we use the similarity score between the landmark representation in the spatial configurations and the object representation in the panoramic images to control the transitions between configurations. Also, we align object representations with the configuration representations enriched with motion indicator, spatial indicator and landmark representations to finally select the navigable image.

A spatial configuration is the smallest linguistic

Proceedings of Second International Combined Workshop on Spatial Language Understanding andGrounded Communication for Robotics, pages 42–52
August 5–6, 2020.

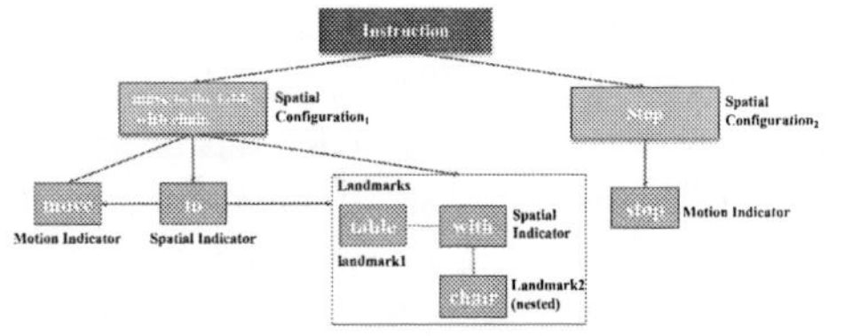

(a) Spatial Configuration Scheme

(b) Spatial Configuration Annotation

Figure 1: **Spatial Configuration example**. The instruction "Move to the table with chair, and stop." can be split into two spatial configurations: "move to the table with chair" and "stop". In configuration1, "move" is motion indicator; "to" is spatial indicator; "table" is landmark. "table with chair" is a nested spatial configuration of configuration1. The role of "table" is trajector; "with" is spatial indicator; and "chair" is landmark. In configuration2, "stop" is motion indicator.

unit that describes the location/trans-location of an object with respect to a reference or a path that can be perceived in the environment. It contains fine-grained spatial roles, such as motion indicator, landmark, spatial indicator, trajector. Essentially, each spatial configuration forms a sub-instruction in our setting. Figure 1 shows an example of splitting an instruction into its corresponding spatial configurations and the extracted spatial roles. Previous research argues representing the semantic structure of the language could improve the reasoning capabilities of deep learning models (Dan et al., 2020; Zheng and Kordjamshidi, 2020). There are relevant work modeling the meaning of spatial semantics in probabilistic models (Kollar et al., 2010; Tellex et al., 2011) and neural models (Regier, 1996; Ghanimifard and Dobnik, 2019). However, its impact on deep learning models for navigation remains an open research problem.

The contribution of this paper is as follows:

1. We consider the spatial semantic structure of the instructions explicitly in terms of spatial configurations and their spatial semantic elements, i.e., spatial/motion indicators, and landmarks to enrich the configuration representations.
2. We introduce a state attention to guarantee that configurations are executed sequentially. Also, we utilize the grounding between the extracted spatial elements and the object representation to help control the transitions between configurations.
3. Our experiment results show that considering the explicit representation of semantic elements of the spatial configurations improves the strong baselines significantly in the seen environments and yields competitive results in the unseen environments.

2 Related Work

Older studies on navigation before the deep learning era are mostly symbolic grounding methods, which are based on parsing the semantics of the instruction and learning probabilistic models. MacMahon et al. (2006) used the parser to associate the linguistic elements in free-form instruction to their corresponding action, location and object in the environment. Tellex et al. (2011) represented the spatial language as a hierarchy of Spatial Description Clauses (SDC) and proposed a discriminative probabilistic graphical model to find the most probable path with the extracted SDC and the detected visual landmark. Mei et al. (2016) provided a good overview of the past classical work on navigation. However, one of the biggest limitations of those methods is that they required prior linguistic structure and manual annotations.

In recent years, given the new capabilities created by deep learning architectures, the navigation task is extended to the photo-realistic simulated environments (Anderson et al., 2018; Thomason et al., 2019; Chen et al., 2019). Based on this, a Sequence-to-Sequence (Seq2seq) baseline model was proposed by Anderson et al. (2018) to encode the instructions and decode the embeddings to identify the corresponding output action sequence with the observed images. Fried et al. (2018) proposed to train a speaker model to augment the instructions for the follower model. Ma et al. (2019) introduced a visual and textual co-attention mechanism and a progress monitor loss to track the execution progress. Although those agents achieved better performance, the semantic structures on both language and vision sides were ignored.

We aim to exploit both symbolic grounding and neural models in the spatial domain. Regier (1996) designed the neurons to learn the meaning of spatial prepositions. Ghanimifard and Dobnik (2019) explored the effects of spatial knowledge in a generative neural language model for the image description. We mainly work on incorporating the spatial semantics in navigation neural agent. Hong et al. (2020) recently provided a method to segment the long instruction into sub-instructions. They used a shifting attention module to infer whether the current sub-instruction has been completed. Sub-instructions differ from us as they manually aligned the instructions and viewpoints to learn the alignments, while we modeled spatial semantics to guide the alignment automatically. Moreover, their proposed shifting attention module is hard attention, and a threshold is set to decide whether the agent should execute the next sub-instruction. However, we utilize the grounding between the landmarks and the objects to control the transitions between sub-instructions.

3 Navigation Model

3.1 Problem Formulation

In this task, the agent follows an instruction to navigate from a start viewpoint to a goal viewpoint in a photo-realistic environment. Formally, the agent is given a natural language instruction S, which is a sequence of tokens, and $\{s_1, s_2, \cdots\}$ is its corresponding token embeddings. The agent observes a 360-degree panoramic view of its surrounding scene at the current viewpoint. Here, we follow Ma et al. (2019) to map the n navigable viewpoints to discrete images from the current panoramic view[1]. We obtain n images corresponding to each navigable viewpoint $I = \{I_1, I_2, \cdots, I_n\}$. The task is to select the next viewpoint among the navigable viewpoints or the current viewpoint (indicating the stop), and finally, to generate the trajectory that takes the agent close to an intended goal location.

3.2 Sequence-to-Sequence

We model the agent with a LSTM-based sequence-to-sequence architecture (Sutskever et al., 2014) to control the flow of information, as illustrated in Fig 2. The encoder computes a contextual embedding $\bar{s}_j$ of each token embedding s_j in S by $\bar{s}_j = LSTM_{encode}(s_j)$. At each step t of navigation, the decoder receives the grounded instruction representation C_t^* and the aligned image representation I_t^* to update its context h_t by $h_t = LSTM_{decode}([C_t^*, I_t^*])$. Finally, we predict the probability distribution of the next navigable viewpoint p_t by h_t. We introduce the method to obtain C_t^* and I_t^* in Section 3.5 and Section 3.6, as well as the next viewpoint prediction in Section 3.7.

3.3 Spatial Configurations Representation

To obtain the configurations in a navigation instruction, we first split the instructions into sentences. Then we design a parser with rules applied on an off-the-shelf dependency parser[2] to extract all the verb phrases and noun phrases in each sentence. In general, each configuration contains at most one motion indicator. Since we aim to process instructions and look for motions, we split the sentences with the extracted verb phrases as motion indicators to obtain spatial configurations. We do not separate the nested configurations with no motion indicator and keep them attached to the dynamic configurations (i.e. the ones with motion-indicator). As shown in Figure 1, "table with chair" is the nested spatial configuration of "move to the table with chair". Here, we only consider the prepositions that are attached to verbs, and merge the spatial indicators and motion indicators such as "move to" and use them together as the motion indicator. After that, we insert a pseudo delimiter token after each configuration and identify their contained noun phrases as landmarks. Each navigation instruction S is split into m configurations. We re-organize the contextual embeddings of tokens $[\bar{s}_1, \bar{s}_2, \cdots]$ generated by the encoder into the array of spatial configurations representation $C = [C_1, C_2 \ldots C_m]$, where m is the number of configurations in the instruction. In the i-th configuration representation $C_i = \left[c_1^i, c_2^i \cdots, c_{\text{P}}^i\right]$, the j-th element c_j^i is the contextual embedding of the corresponding k-th tokens in the instruction: $c_j^i = \bar{s}_k$. The last token of each configuration is always the pseudo delimiter indexed by P, which contains the most comprehensive context information about the preceding words. Soft attention is widely used to merge a collection of representations V into one by weighted sum based on the relevance indicated by their associated keys representations K and a query Q, calculated by Eq. 1.

$$\text{SoftAttn}(Q; K; V) = \text{softmax}\left(\frac{Q^T W K}{\sqrt{d_k}}\right) V \quad (1)$$

where W is a trainable linear mapping, and d_k is the dimension of each representation in K. We apply a soft attention to each configuration representation with the pseudo delimiter representation c_{P}^i, which can be calculated by Eq. 2.

$$\bar{C}_i = \text{SoftAttn}_{\text{config}}(Q = c_{\text{P}}^i; K = C_i; V = C_i) \quad (2)$$

After obtaining configuration representations, an agent needs to identify which configuration to fol-

[1] 12 headings and 3 elevations with 30 degree interval.

[2] https://spacy.io/

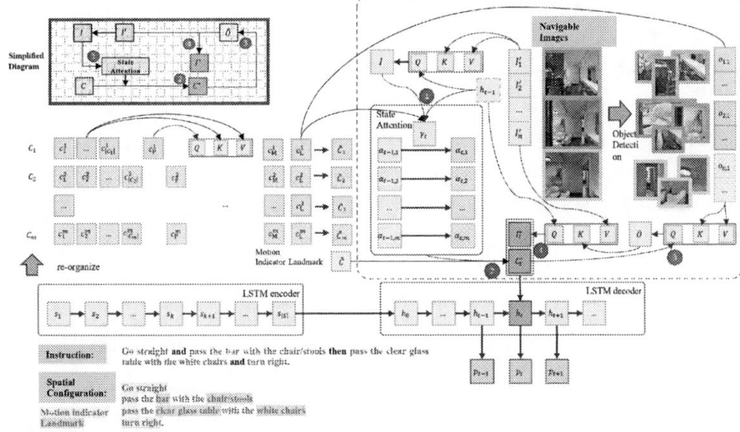

Figure 2: **Model Architecture.** The input to the encoder is the instruction text. The inputs to the decoder are the grounded language C_t^* calculated by state attention and the aligned visual representations I_t^* obtained from navigable images at each step t. The decoder predicts the distribution of next viewpoint p_t with the updated context h_t. The high-level view at the top-left shows the information flow in the model aligning with the circled numbers.

low at each step. To achieve this, we incorporate the intra-configuration and inter-configuration knowledge. Concretely, intra-configuration knowledge is the motion indicator that guides the agent movement and the landmarks that could be grounded into the objects in visual images; inter-configuration knowledge is that configurations should be processed one after another.

As mentioned above, we identify verbs and noun chunks in configurations as motion indicator and landmarks respectively. Each configuration can contain only one motion indicator and multiple landmarks. Formally, for the i-th configuration C_i, the motion indicator representation is denoted as c_{M}^i and the landmark representation is denoted as $c_L^i = \left[c_{\mathrm{L}_1}^i, c_{\mathrm{L}_2}^i, \cdots, c_{\mathrm{L}_p}^i\right]$, where p is the number of landmarks. If there is no landmark in the configuration, the value of c_{L}^i will be set as zeros. To enhance the motion indicator and landmark information, we concatenate their word embedding with the configuration representation. In case there are multiple noun chunks in configuration, to simplify, we select the noun closest to the root of the parsing tree as the main landmark, denoted as $\hat{p}$. Then the enriched configuration representation is denoted as $\tilde{C}_i = \left[\bar{C}_i; c_{\mathrm{M}}^i; c_{\mathrm{L}_{\hat{p}}}^i\right]$.

3.4 Visual Representation

To execute a series of configurations, the agent needs to keep track of the sequence of images observed along the navigation trajectory. We firstly transform the low-level image features from ResNet of n navigable images $I = \{I_1, I_2, \ldots, I_n\}$ to $I' = \left[I_1', I_2', \cdots, I_n'\right]$ by a fully-connected layer $I_j' = \mathrm{FC}_{\mathrm{img}}(I_j)$. Then, a soft attention is applied to I' with the previous context h_{t-1}, as shown in Eq. 3.

$$\bar{I} = \mathrm{SoftAttn}_{\mathrm{img}}(Q = h_{t-1}; K = I'; V = I') \tag{3}$$

Furthermore, we equip the agent with object-based representation. Specifically, we get top-K object representations from each image with an object detection model[3]. In this paper, we consider two kinds of object representation: object label representation and object visual representation. Specifically, the label representation uses the GloVe embedding (Pennington et al., 2014) of the type of the object, and visual representation uses the region-of-interest (ROI) pooling of the object detection model. We will compare the two representations and a hybrid representation of them in Appendix A.1. Formally, the object representations could be denoted as $O = [O_1, O_2 \ldots O_n]$, where for image I_j, there is $O_j = [o_{j,1}, o_{j,2}, \cdots, o_{j,K}]$. $o_{j,k}$ is the k-th object representation in j-th image.

3.5 Spatial Configuration Grounding

To guarantee the sequential execution, we design a state attention mechanism over the configurations.

[3]We employ Faster R-CNN pre-trained on Visual Genome, and use at most 36 objects that have an area greater than 10 pixels.

We consider the attention weight at each step as a state that measures navigation progress and is updated by a controller. Formally, the i-th configuration at step t is denoted as $\alpha_{t,i}$. At the first step, the attention weight is initialized to be focused on the first configuration $\alpha_0 = [1, 0, \cdots]$. At each of the following steps, the attention weight is updated by a controller γ_t with discrete convolution. γ_t is a two dimensional probability distribution indicating to what extent the agent should execute the current configuration or move to the next. The updating process is formally defined in Eq. 4.

$$\alpha_{t,i} = \sum_{\imath=i-1}^{i} \alpha_{t-1,\imath} \cdot \gamma_{t,i-\imath} \tag{4}$$

Using a set of rules to determine the value of the controller γ is not practical. For example, for the instruction "move to the table" or "move past the table", it is hard for an agent to decide whether to execute the current configuration or to move to the next one only based observing or not-observing the "table". To address this issue, we let the agent learn the value of γ based on three aspects of information. The first one is the previous hidden state h_{t-1}; the second one is the attended image representation $\bar{I}_t$ at the current step; the third one is the similarity score S_t between the landmark representations and the object representations, Eq. 5 shows how to get the similarity score S_t, and α_{t-1} is the attention weight at the previous step.

$$S_t = \tilde{C}_L \cdot O \cdot \alpha_{t-1} \tag{5}$$

Then, we use a fully connected layer to predict the distribution $\gamma_t = \text{FC}_\gamma\left(\left[h_{t-1}; \bar{I}_t; S_t\right]\right)$. Finally, we apply the state attention to $\tilde{C}$ to get the grounded instruction representation based on the configuration $\hat{C} = \sum_i \alpha_{t,i} \cdot \tilde{C}_i$, which is used as the language input to the decoder $C_t^* = \hat{C}$.

3.6 Visual Representation Alignment

The intuition to leverage the object representation is to select navigable images by aligning the object representation with the configuration representation. We use two levels of soft attention, first over the objects in each image by configuration representation $\hat{C}$, and second over all images guided by the previous context h_{t-1}.

$$\begin{aligned} \hat{O}_j &= \text{SoftAttn}_{\text{obj}}(Q = \hat{C}; K = O_j; V = O_j) \\ \hat{I} &= \text{SoftAttn}_{\text{objimg}}(Q = h_{t-1}; K = \hat{O}; V = I') \end{aligned} \tag{6}$$

where $\hat{O} = \left[\hat{O}_1, \hat{O}_2, \cdots, \hat{O}_n\right]$. We use the image representation $\hat{I}$, that has aligned the objects with the configurations, as the visual input to the decoder $I_t^* = \hat{I}$.

3.7 Navigable Viewpoint Selection

We obtain a new decoder context h_t, as described in Section 3.2, with configuration input C_t^* and visual input I_t^*, where t is the current step. The next step is to predict the viewpoint with the image that has the highest correlation with the current context and configuration, calculated by $z_{t,j} = \left\langle I_j', \text{FC}_{\text{pred}}\left([C_t^*; h_t]\right)\right\rangle$, where $\text{FC}_{\text{pred}}(\cdot)$ is a fully-connected layer. We sum the scores of the three elevations for each navigable viewpoint k as $\zeta_{t,k} = \sum_{j \in \kappa_k} z_{t,j}$, where κ_k is the set of three elevations' image indexes. The predicted navigable viewpoint distribution p_t can be calculated with $p_t = \text{softmax}(\zeta_t)$.

3.8 Training and Inference

We train our model with two state-of-the-art training strategies in this task. (1) **T1**: We follow Self-Monitor (Ma et al., 2019) optimizing the model with a cross-entropy loss to maximize the likelihood of the ground-truth navigable viewpoint given by the model, and a mean squared error loss to minimize the normalized distance in units of length from the current viewpoint to the goal destination. At each step, the next viewpoint is selected by sampling the predicted probability of each navigable viewpoint. (2) **T2**: We follow (Tan et al., 2019) training the model with the mixture of Imitation Learning and Reinforcement Learning, where Imitation Learning minimizes the cross-entropy loss of the prediction and always samples the ground-truth navigable viewpoint at each time step, and Reinforcement Learning uses policy gradient to update the parameters of the model.

During inference, we conduct a greedy search with the highest probability of the next viewpoints to generate the trajectory. It should be noticed that beam search with a beam size greater than one is not practical because the agent needs to move forward and backward in the physical world, resulting in a long trail trajectory before making a decision.

4 Experimental Setup

Dataset We evaluate our model with Room-to-Room (R2R) dataset (Anderson et al., 2018), which is built upon the Matterport3D dataset (Chang et al.,

	Method	Validation-Seen NE ↓	**SR ↑**	**SPL ↑**	Validation-Unseen NE ↓	**SR ↑**	**SPL ↑**	Test(Unseen) NE ↓	**SR ↑**	**SPL ↑**
1	Random (Anderson et al., 2018)	9.45	0.16	-	9.23	0.16	-	9.77	0.13	0.12
2	Student-forcing (Anderson et al., 2018)	6.01	0.39	-	7.81	0.22	-	7.85	0.20	0.18
3	Speaker-Follower (Fried et al., 2018)	4.36	0.54	-	7.22	0.27	-	-	-	-
4	Speaker-Follower*	3.66	0.66	0.58	6.62	0.36	-	6.62	0.35	0.28
5	Self-Monitor* (Ma et al., 2019)	**3.22**	**0.67**	0.58	5.52	0.45	0.32	**5.67**	0.48	0.35
6	Environment Dropout* (Tan et al., 2019)	4.19	0.58	0.55	**5.43**	**0.48**	**0.44**	-	**0.52**	**0.47**
7	Environment Dropout + BERT*	4.40	0.61	0.57	5.54	0.46	0.43	-	-	-
8	SpC-NAV*	4.09	0.65	**0.61**	5.92	0.45	0.42	6.22	0.46	0.44

Table 1: **Experimental Result comparing with baseline models.** * means data augmentation.

Method	Val-Seen NE ↓	**SR ↑**	**SPL ↑**	Val-Unseen NE ↓	**SR ↑**	**SPL ↑**	Test(Unseen) NE ↓	**SR ↑**	**SPL ↑**
Self-Monitor (T1)	3.72	0.63	0.56	5.98	0.44	0.30	-	-	-
Sub-Instruction(T1)	-	-	-	**6.16**	**0.42**	**0.32**	-	-	-
SpC-NAV+T1	3.95	0.65	0.59	6.51	0.39	**0.32**	6.22	0.42	0.35
EnvDrop (T2)	4.71	0.55	0.53	**5.49**	**0.47**	**0.43**	-	-	-
Sub-Instruction(T2)	-	-	-	5.67	0.47	**0.43**	-	-	-
SpC-NAV+T2	**4.68**	**0.59**	**0.56**	6.68	0.44	0.39	6.25	0.45	0.43

Table 2: **Experimental Result with Different Training Strategies**. T1 and T2 are two training strategies.

2017). This dataset has 7,189 paths and 21,567 instructions with an average length of 29 words. The whole dataset is divided into training, seen validation, unseen validation, and (unseen) test sets. The seen validation set shares the same visual environments with the training set, while unseen validation and test sets contain different environments.

Evaluation Metrics We report three evaluation metrics. (1) Navigation Error (NE): the mean of the shortest path distance between the agent's final position and the goal location. (2) Success Rate (SR): the percentage of the cases where the predicted final position lays within 3m from the goal location. (3) Success rate weighted by normalized inverse Path Length (SPL): SPL normalize Success Rate by trajectory length (Anderson et al., 2018). SPL is recommended as the primary metric because it considers both the effectiveness and efficiency of navigation performance.

4.1 Baseline Models

We mainly compare Spc-NAV with the following baseline models. **Seq2Seq** (Anderson et al., 2018) trained an encoder-decoder model with two learning strategies of random and student-forcing. **Speaker-Follower** (Fried et al., 2018) introduced a speaker module to synthesize new instructions to train the follower module. **Self-Monitor** (Ma et al., 2019) co-grounded instructions and image based on soft attention mechanism. **Environmental Dropout** (Tan et al., 2019) proposed a neural agent trained with the method of the mixture of Imitation Learning and Reinforcement Learning. **Sub-instruction** (Hong et al., 2020) segmented the instruction into sub-instructions and designed a shifting attention module to ensure the sequential execution order between sub-instructions. The differences between Sub-instruction and our model has been discussed in Section 2.

4.2 Implementation Details

We implement SpC-NAV using PyTorch [4] We use 768-d BERT-base (Devlin et al., 2018) (frozen) as the embedding of the raw instruction, and get its 512-d contextual embedding by LSTM. We encode the representations of the motion indicator and the landmark in each configuration with 300-d GloVe embedding respectively, and concatenate them with the 512-d configuration representation to obtain the enriched configuration representation (1112-d). We use 300-d GloVe embedding of object label representation to calculate similarity score S with configuration representation. We trained an auto-encoder to map 2048-d object visual representation from Faster R-CNN to 152-d, and use it to obtain the attended object representation $\hat{O}$. We optimize using ADAM with learning rate $1e-4$ in batches of 64. We used a rule-based parser to obtain the spatial configuration and spatial semantic elements. This provides some noisy extractions. Appendix A.2 includes the details about the accuracy of the parser based on our manual annotations of a subset of instructions.

5 Results and Analysis

Table 1 shows the main performance metrics of our proposed SpC-NAV, compared with the baseline models on seen/unseen validation set and unseen testing set. To achieve the best result, SpC-NAV is trained with the training strategy T2 (see Section 3.8) and the data augmentation proposed in (Tan et al., 2019). Our model improves the performance in the seen environment and obtains com-

[4]https://github.com/zhangyuejoslin/SpC-NAV

petitive results in the unseen environment. Since we use BERT as the input to the encoder while the baseline models use basic word embeddings, we replace the word representations in Environment Dropout with BERT for a fair comparison. Although the richer language representations help the performance, our model still achieves better results, especially in the seen environments. It indicates that the spatial configuration and spatial elements indeed improve the agent's reasoning ability.

Training strategies are orthogonal to our work, and our model is friendly to the strategies widely used in the literature (T1/T2) (see Section 3.8). We evaluate SpC-NAV with both T1 and T2 and compare the results with their baseline models as well as Sub-Instruction. We do not apply data augmentation in this setting. As shown in Table 2, SpC-NAV achieves consistent improvement in the seen environment compared with all the baselines. In the unseen environment, training with T1, SpC-NAV outperforms Self-monitor (and is even comparable to it with data augmentation) and performs similarly as Sub-Instruction. However, training with T2, our model does not outperform Environment Dropout and Sub-Instruction in unseen environments. We analyze the errors in Section 5.2.

5.1 Ablation Analysis

Table 3 shows how various spatial semantic elements influence the performance of the model. The model is trained with the training strategy T1. Row#1 is our model without considering spatial elements. From row#2 to row#3, we incorporate the representations of the motion indicator and the landmark into spatial configuration representation incrementally. In row#4, we use the similarity score between the landmark representations in the configuration and the object label representations in the image to control the transitions between spatial configurations. All motion indicator, landmark and similarity score improve the performance. After applying the similarity score, the large gain indicates that the connection between landmarks and objects is important in language grounding.

5.2 Qualitative Analysis

Seen Environment

We analyze some qualitative examples to find out how the spatial semantics improve the model. For the semantics of motion, we find that our model can improve the cases that motions contain "up" and "down" after adding the representation of motion indicator. Figure 3 (a) shows an example of such a scenario. The spatial configuration is "walk up the stairs", and the agent could find the right viewpoints after we incorporated the representation of the motion indicator "walk up". However, the model makes more mistakes in the cases that the motion indicators are highly related to the objects, such as "walk through", "walk past", and "walk towards", which need the landmark information. In these latter cases, the model should consider both motions and landmarks together. In another experiment, we added the landmark representation. Figure 3 (b) shows an example that the spatial configurations is "walk past the dining room table". The agent can select the correct viewpoints when we incorporate the representation of landmark "dining room table". We also analyze the influence of the similarity score, and found that when the information in the current configuration is not sufficient to make a decision, the similarity score will assist in choosing the next configuration. For example, in Figure 3 (c), the spatial configurations are "turn right" and "walk past the couch". Without using the similarity score in controlling the transitions between configurations, the agent tends to select a viewpoint in the "right" direction. But with similarity score, the agent will consider both "turn right" and "walk past the couch", and selects the correct viewpoint that the "couch" can be seen.

	Validation-Seen			Validation-Unseen		
Model	NE↓	**SR↑**	**SPL↑**	NE↓	**SR↑**	**SPL↑**
1 SpC-NAV	4.11	0.62	0.53	6.49	0.39	0.29
2 SpC-NAV$_{M}$	**3.88**	0.62	0.53	**6.21**	**0.40**	0.28
3 SpC-NAV$_{M+L}$	4.01	0.62	0.54	6.27	0.39	0.29
4 SpC-NAV$_{M+L+S}$	3.95	**0.65**	**0.59**	6.51	0.39	**0.32**

Table 3: **Ablation study with different spatial semantics.** The subscription letters mean the model took those information into account; *M: motion indicator; L: landmark; S: similarity score.*

Unseen Environment

Table 1 and Table 2 show that our model does not outperform Environment Dropout in the unseen environments. We noticed that the main error is that some objects can not be detected in the image by the object detection model. This is more problematic for our model because we explicitly align the landmark phrases with the detected objects. For example, in Fig 4 (a), the agent selects the correct viewpoint when the configuration is "Walk to the glass door" because the connection between the landmark "glass door" and the object "door" has

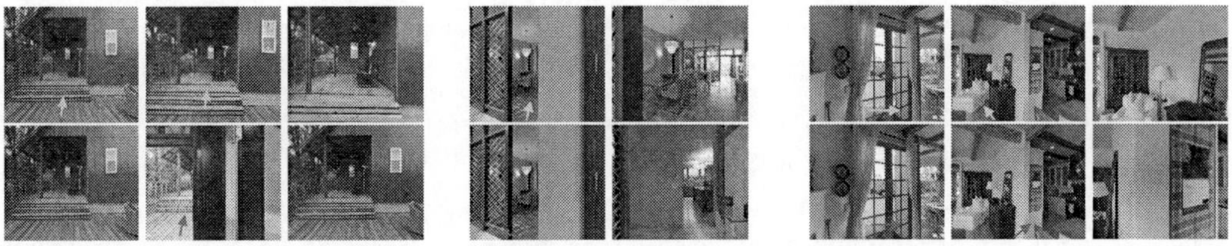

(a) ±Motion Indicator
Walk up the stairs.

(b) ±Landmark
Walk past the dining room table.

(c) ±Similarity Score
Turn right, and walk past the couch.

Figure 3: **Analysis of Seen examples.** In these three scenarios, the corresponding spatial configurations are provided. Green arrows in the above figures show the correct trajectory was selected after the additional spatial semantics; red arrows show without that information the agent went wrong.

(a) Walk to the **glass door**.

(b) Go to the **pottery**.

Figure 4: **Analysis of an Unseen example**

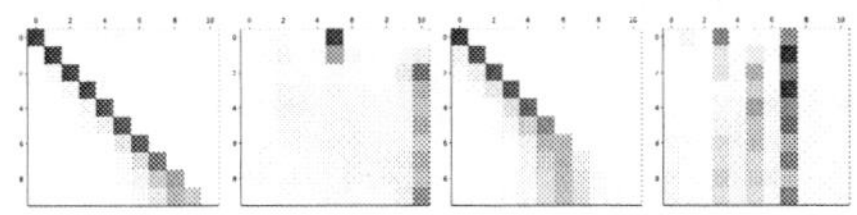

(a) State Seen (b) Soft Seen (c) State Unseen (d) Soft Unseen

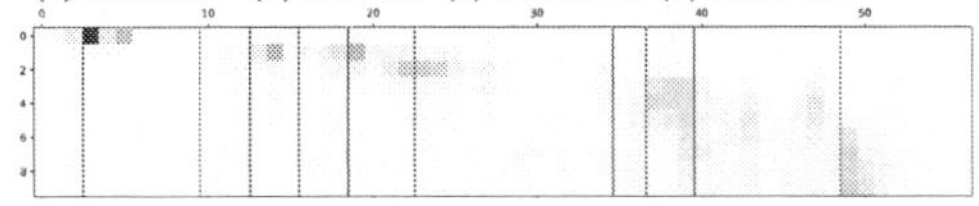

(e) Soft seen of Self-Monitor

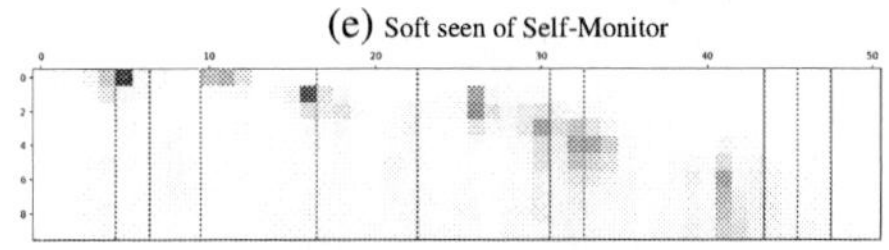

(f) Soft unseen of Self-Monitor

Figure 5: **Attention weights of various attention strategies with seen and unseen examples.** The horizontal axis is the configuration order, and the vertical axis is the temporal order of the steps taken by the agent. Each row in sub-figures show the attention distribution over the configurations (or tokens) in an instruction at each time step. The green vertical lines in Figure (e) and Figure (f) indicate the split points of the configurations in the instruction.

been learned in training set. In Fig (b), the agent is wrong when the configuration is "Go to the pottery." because the "pottery" is not detected at the initial perspective and the word "pottery" never appears in the training set. However, the agent selects a viewpoint that a bounding box contains a pottery. The gap between seen and unseen become larger after data augmentation since our model is able to capture the structure of the language by observing more examples. It can deal with the variations in the instructions and improve the performance in the seen environment, but it fails to deal with the novel objects and visual variations in the unseen environments. This is an orthogonal issue addressed in zero-shot learning (Blukis et al., 2020).

5.3 State Attention Visualization

We visualize the state attention and the soft attention weights over configurations. As shown in Fig 5a and Fig 5c, our designed state attention demonstrates that the grounded configuration shifts gradually from the first configuration to the last in both seen and unseen environments. We apply the soft attention used in Self-Monitor on spatial configurations, as shown in Fig 5b and Fig 5d, it can not preserve the sequential execution order. We also show the soft attention weights of the grounded instruction in the Self-Monitor by splitting the instructions with the boundaries of our configurations. As shown in Fig 5e and Fig 5f, although their attention weights show the gradual shift, many configurations are skipped.

6 Conclusion

We propose a neural agent that incorporates the semantic elements of spatial language for vision-and-language navigation. We use the notion of spatial configurations as the main linguistic unit of the instructions and enhance the spatial configuration representation with the representations of motion indicator and landmark. We design a state attention to guarantee the sequential execution order of configurations and use the similarity score between the representations of landmarks and objects to control the transitions between configurations. Based on our results, incorporating the spatial semantics improves reasoning ability over navigation. Future work could investigate more fine-grained spatial semantics and the geometry of spatial relations. Also, we will deal with novel objects in a zero-shot setting to improve the unseen environments results.

Acknowledgement

This project is supported by National Science Foundation (NSF) CAREER award 2028626 and partially supported by the Office of Naval Research (ONR) grant N00014-20-1-2005. Any opinions, findings, and conclusions or recommendations expressed in this material are those of the authors and do not necessarily reflect the views of the National Science Foundation nor the Office of Naval Research. We thank all reviewers for their thoughtful comments and suggestions.

References

Peter Anderson, Qi Wu, Damien Teney, Jake Bruce, Mark Johnson, Niko Sünderhauf, Ian Reid, Stephen Gould, and Anton van den Hengel. 2018. Vision-and-language navigation: Interpreting visually-grounded navigation instructions in real environments. In *Proceedings of the IEEE Conference on Computer Vision and Pattern Recognition*, pages 3674–3683.

Valts Blukis, Ross A Knepper, and Yoav Artzi. 2020. Few-shot object grounding and mapping for natural language robot instruction following. *arXiv preprint arXiv:2011.07384*.

Angel Chang, Angela Dai, Thomas Funkhouser, Maciej Halber, Matthias Niessner, Manolis Savva, Shuran Song, Andy Zeng, and Yinda Zhang. 2017. Matterport3d: Learning from rgb-d data in indoor environments. *arXiv preprint arXiv:1709.06158*.

Howard Chen, Alane Suhr, Dipendra Misra, Noah Snavely, and Yoav Artzi. 2019. Touchdown: Natural language navigation and spatial reasoning in visual street environments. In *Proceedings of the IEEE Conference on Computer Vision and Pattern Recognition*, pages 12538–12547.

Soham Dan, Parisa Kordjamshidi, Julia Bonn, Archna Bhatia, Zheng Cai, Martha Palmer, and Dan Roth. 2020. From spatial relations to spatial configurations. In *Proceedings of The 12th Language Resources and Evaluation Conference*, pages 5855–5864.

Jacob Devlin, Ming-Wei Chang, Kenton Lee, and Kristina Toutanova. 2018. Bert: Pre-training of deep bidirectional transformers for language understanding. *arXiv preprint arXiv:1810.04805*.

Daniel Fried, Ronghang Hu, Volkan Cirik, Anna Rohrbach, Jacob Andreas, Louis-Philippe Morency, Taylor Berg-Kirkpatrick, Kate Saenko, Dan Klein, and Trevor Darrell. 2018. Speaker-follower models for vision-and-language navigation. In *Advances in Neural Information Processing Systems*, pages 3314–3325.

Mehdi Ghanimifard and Simon Dobnik. 2019. What goes into a word: generating image descriptions with top-down spatial knowledge. In *Proceedings of the 12th International Conference on Natural Language Generation*, pages 540–551.

Yicong Hong, Cristian Rodriguez-Opazo, Qi Wu, and Stephen Gould. 2020. Sub-instruction aware vision-and-language navigation. *arXiv preprint arXiv:2004.02707*.

Ronghang Hu, Daniel Fried, Anna Rohrbach, Dan Klein, Kate Saenko, et al. 2019. Are you looking? grounding to multiple modalities in vision-and-language navigation. *arXiv preprint arXiv:1906.00347*.

Thomas Kollar, Stefanie Tellex, Deb Roy, and Nicholas Roy. 2010. Toward understanding natural language directions. In *2010 5th ACM/IEEE International Conference on Human-Robot Interaction (HRI)*, pages 259–266. IEEE.

Chih-Yao Ma, Jiasen Lu, Zuxuan Wu, Ghassan AlRegib, Zsolt Kira, Richard Socher, and Caiming Xiong. 2019. Self-monitoring navigation agent via auxiliary progress estimation. *arXiv preprint arXiv:1901.03035*.

Matt MacMahon, Brian Stankiewicz, and Benjamin Kuipers. 2006. Walk the talk: Connecting language, knowledge, and action in route instructions. *Def*, 2(6):4.

Hongyuan Mei, Mohit Bansal, and Matthew R Walter. 2016. Listen, attend, and walk: Neural mapping of navigational instructions to action sequences. In *Thirtieth AAAI Conference on Artificial Intelligence*.

Jeffrey Pennington, Richard Socher, and Christopher D Manning. 2014. Glove: Global vectors for word representation. In *Proceedings of the 2014 conference on empirical methods in natural language processing (EMNLP)*, pages 1532–1543.

Terry Regier. 1996. *The human semantic potential: Spatial language and constrained connectionism*. MIT Press.

Ilya Sutskever, Oriol Vinyals, and Quoc V Le. 2014. Sequence to sequence learning with neural networks. In *Advances in neural information processing systems*, pages 3104–3112.

Hao Tan, Licheng Yu, and Mohit Bansal. 2019. Learning to navigate unseen environments: Back translation with environmental dropout. *arXiv preprint arXiv:1904.04195*.

Stefanie Tellex, Thomas Kollar, Steven Dickerson, Matthew R Walter, Ashis Gopal Banerjee, Seth Teller, and Nicholas Roy. 2011. Understanding natural language commands for robotic navigation and mobile manipulation. In *Twenty-fifth AAAI conference on artificial intelligence*.

Jesse Thomason, Michael Murray, Maya Cakmak, and Luke Zettlemoyer. 2019. Vision-and-dialog navigation. *arXiv preprint arXiv:1907.04957*.

Xin Wang, Qiuyuan Huang, Asli Celikyilmaz, Jianfeng Gao, Dinghan Shen, Yuan-Fang Wang, William Yang Wang, and Lei Zhang. 2019. Reinforced cross-modal matching and self-supervised imitation learning for vision-language navigation. In *Proceedings of the IEEE Conference on Computer Vision and Pattern Recognition*, pages 6629–6638.

Haonan Yu, Xiaochen Lian, Haichao Zhang, and Wei Xu. 2018. Guided feature transformation (gft): A neural language grounding module for embodied agents. *arXiv preprint arXiv:1805.08329*.

Chen Zheng and Parisa Kordjamshidi. 2020. Srl-grn: Semantic role labeling graph reasoning network. *arXiv preprint arXiv:2010.03604*.

Wang Zhu, Hexiang Hu, Jiacheng Chen, Zhiwei Deng, Vihan Jain, Eugene Ie, and Fei Sha. 2020. Babywalk: Going farther in vision-and-language navigation by taking baby steps. *arXiv preprint arXiv:2005.04625*.

A Appendix

A.1 Visual Representation Analysis

In this section, we experiment with three types of object representations introduced in Section 3.6, which are object label representation and object visual representation and the combination of these two types of object representation. As shown in Table 4, object visual representation performs better in unseen environments, and we use it to get attended object representation $\hat{O}$ in our best model. This experiment does not consider the similarity score between the representations of landmarks and objects.

	Validation-Seen			Validation-Unseen		
Repr.	NE↓	**SR↑**	**SPL↑**	NE↓	**SR↑**	**SPL↑**
Label	4.51	0.58	0.52	6.43	0.37	0.28
Visual	4.01	0.62	0.54	6.27	0.39	0.29
Label + Visual	4.45	0.59	0.53	6.54	0.37	0.28

Table 4: Result with Different Visual Representations.

A.2 Parsing Analysis

The performance of our rule-based parser influences the result of navigation. To evaluate it, we manually annotated 845 spatial configurations for 200 instructions. We annotated motion indicators, spatial indicators and landmarks in those configurations. Our parser achieves an accuracy of 85% in extracting the spatial configurations. For the extraction of spatial elements, the accuracy is 73% for motion/spatial indicators, and 77% for landmarks.

In the following, we analyze two types of error in getting spatial configurations (Split Error and Order Error), and other errors that generated in the extraction of motion indicator, spatial indicator and landmark.

Split Error

The split configuration may only convey the spatial position of objects rather than executable navigation information. For example, in the instruction, "Turn left. There is a rocking chair in it," two configurations are generated based on our split method: "Turn left" and "There is a rocking chair in it." However, the second configuration is not an independent spatial configuration because it indicates no motion, and it is attached to the previous configuration.

Order Error

We order the configurations based on their occurrence in the sentence. However, there are cases that the configurations have an inverted order. For instance, "Stop once you pass the counter on the right" is split as "stop" and "you pass the counter on the right." However, the implied sequence is inverted because of "once".

Motion Indicator and Spatial Indicator

We build a vocabulary based on training data to collect the commonly used verb phrases, and the vocabulary size is 241. Table 5 shows some examples. If the motion indicator and spatial indicator does not show in the vocabulary, we will treat the verbs as the motion indicators and prepositions as spatial indicators in configurations. With this method, we can get 73% accuracy since there are expressions that never appear in the training dataset, and it is hard to extract the complete verb phrases only based on pos-tag.

Landmark

We extract the noun phrases of each configuration as landmark and can get 77% accuracy. However, there are some special cases, for example, "a left" in "make a left" is extracted as noun chunk, but it can not be treated as a landmark. Also, for the expression "middle of the doorway", "the middle" and "the doorway" are both noun chunks, but the whole phrase is the landmark instead of separated ones.

head straight, walk through, walk down, walk into, walk inside, turn around, turn left, make a left turn, jump over, move forward, turn slightly right

Table 5: Verb Phrases Examples

Error-Aware Interactive Semantic Parsing of OpenStreetMap

Michael Staniek
Computational Linguistics
Heidelberg University
Germany
staniek@cl.uni-heidelberg.de

Stefan Riezler
Computational Linguistics & IWR
Heidelberg University
Germany
riezler@cl.uni-heidelberg.de

Abstract

In semantic parsing of geographical queries against real-world databases such as OpenStreetMap (OSM), unique correct answers do not necessarily exist. Instead, the truth might be lying in the eye of the user, who needs to enter an interactive setup where ambiguities can be resolved and parsing mistakes can be corrected. Our work presents an approach to interactive semantic parsing where an explicit error detection is performed, and a clarification question is generated that pinpoints the suspected source of ambiguity or error and communicates it to the human user. Our experimental results show that a combination of entropy-based uncertainty detection and beam search, together with multi-source training on clarification question, initial parse, and user answer, results in improvements of 1.2% F1 score on a parser that already performs at 90.26% on the NLMaps dataset for OSM semantic parsing.

1 Introduction

Semantic Parsing has the goal of mapping natural language questions into formal representations that can be executed against a database. If real-world large-scale databases such as OpenStreetMap (OSM)[1] need to be accessed, the creation of gold standard parses by humans can be complicated and requires expert knowledge, and even reinforcement learning from answers might be impossible since unique correct answers to OSM queries do not necessarily exist. Instead, uncertainties can arise due to open-ended lists (e.g., of restaurants), fuzzily defined geo-positional objects (e.g., objects "near" or "in walking distance" of other objects), or by ambiguous mappings of natural language to OSM tags[2], with the truth lying in the eye of the beholder who asked the original question. Semantic parsing against OSM thus asks for an interactive setup where an end-user inter-operates with a semantic parsing system in order to negotiate a correct answer, or to resolve parsing ambiguities and to correct parsing mistakes, in a dialogical process.

Previous work on interactive semantic parsing (Labutov et al., 2018; Yao et al., 2019; Elgohary et al., 2020) has put forward the following dialogue structure: i) the user poses a natural language question to the system, ii) the system parses the user question and explains or visualizes the parse to the user, iii) the user generates natural language feedback, iv) the parser tries to utilize the user feedback to improve the parse of the original user question. In most cases, the "explanation" produced by the system is restricted to a rule-based reformulation of the parse in a human intelligible form, whereas the human user has to take guesses about where the parse went wrong or is ambiguous.

The goal of our paper is to add an explicit step of error detection on the parser side, resulting in an automatically produced clarification question that pinpoints the suspected source of ambiguity or error and communicates it to the human user. Our experimental results show that a combination of entropy-based uncertainty detection and beam search for differences to the top parse yield concise clarification questions. We create a dataset of 15k clarification questions that are answered by extracting information from gold standard parses, and complement this with a dataset of 960 examples where human users answer the automatically generated questions. Supervised training of a multi-source neural network that adds clarification questions, initial parses, and user answers to the input results in improvements of 1.2% F1 score on

[1] www.openstreetmap.org

[2] For example, *recreation grounds* can map to tags reserved for *leisure* purposes or for official *landuse* registration; *bars* map to tags *bar* and *pub* that differ in that only the latter sells food; *off-license* shops can have licenses to sell only *wine* or all kinds of *alcohol*.

Proceedings of Second International Combined Workshop on Spatial Language Understanding andGrounded Communication for Robotics, pages 53–59
August 5–6, 2020.

a parser that already performs at 90.26% on the NLMaps dataset for OSM semantic parsing.

2 Related Work

Yao et al. (2019) interpret interactive semantic parsing as a slot filling task, and present a hierarchical reinforcement learning model to learn which slots to fill in which order. They claim the automatic production of clarification questions by the agent as a main feature of their approach, however, what is actually used in their work is a set of 4 predefined templates. Elgohary et al. (2020) show an interpretation of the parse that is understandable for laypeople with a template-based approach, and present different approaches to utilize the user response to improve the parser. In their work, the explantion on the parser side is purely template-based, whereas our work explicitly informs the clarification question by possible sources of parse ambiguities or errors.

Considerable effort has been invested in the creation of large datasets for parsing into SQL representations. Yu et al. (2018) created a dataset called Spider which is a complex, cross-domain semantic parsing and text-to-SQL dataset. Their annotation process was very extensive, and involved 11 computer science students who invested a total of 1,000 hours into asking natural language queries and creating the corresponding SQL query. Extensions of the Spider dataset, SParC (Yu et al., 2019b), or Co-SQL (Yu et al., 2019a) involved even more computer science students. Our work attempts an automatic construction of concise clarification questions, allowing for faster dataset construction.

3 (Multi-Source) Neural Machine Translation

Our work employs as a semantic parser a sequence-to-sequence neural network (Sutskever et al., 2014) that is based on an recurrent encoder and decoder architecture with attention (Bahdanau et al., 2015).

Given a corpus of aligned data $D = \{(x_n, y_n)\}_{n=1}^{N}$ of user queries x and semantic parses y, standard supervised training is performed by minimizing a Cross-Entropy objective $-\frac{1}{N}\sum_{n=1}^{N}\sum_{t=1}^{T}\log p(y_{n,t}|y_{n,<t},x_n)$, where the probability of the full output sequence $y = y_1, y_2, ..., y_n$ is calculated by the product of the probability for every timestep where $p(y|x) = \prod_{t=1}^{T} p(y_t|y_{<t}, x)$.

This model can be easily extended to multi-source learning (Zoph and Knight, 2016) by using not only one, but multiple encoders. This means that there are actually multiple sequences of hidden states. The decoder hidden state is consequently initialized by a linear projection of the average of the last hidden states of all encoders $c = \frac{1}{N}\sum_{i=1}^{N} h_i W_l$, and needs to implement a separate attention mechanism for every encoder.

To be able to fine-tune a model with feedback from a user, the standard cross-entropy objective cannot be used because the desired target is not a gold parse, but a parse $\tilde{y}$ predicted by the system, that has been annotated with positive and negative markings by a human user. This can be formalized as assigning a reward δ_t that is either positive or negative to every token in the parse ($\delta_{t+} = 0.5$ and $\delta_{t-} = -0.5$). It is then possible to maximize the likelihood of the correct parts of the parse by optimizing a weighted supervised learning objective $\sum_{x,\tilde{y}}\sum_{t=1}^{T}\delta_t \log p(\tilde{y}_t|x, y_{<t})$. (Petrushkov et al., 2018)

4 Neural Semantic Parsing of OSM

4.1 Data

Our work is based on the NLmaps v2 dataset.[3] NLmaps builds on the Overpass API which allows the querying of the OSM database with natural language queries. This dataset includes template-based expansions leading to duplicates in train and test sets. However, these expansions introduced problematic features into the data in that OSM tags were inserted which, according to the documentation in the OSM developer wiki, should not be used:

- Is there Recreation Grounds in Marseille

```
→ query(area(keyval('name','Marseille')),
  nwr(keyval('leisure','recreation_ground'),
  qtype(least(topx(1))))
```

- Recreation Ground in Frankfurt am Main

```
→ query(area(keyval('name','Frankfurt
  am Main')),
  nwr(keyval('landuse','recreation_ground')),
  qtype(latlong))
```

While *leisure=recreation_ground* certainly exists as a tag[4], its use is heavily discouraged[5]. Furthermore, several mistakes were introduced in the

[3] www.cl.uni-heidelberg.de/statnlpgroup/nlmaps/

[4] https://wiki.openstreetmap.org/wiki/Tag:leisure%3Drecreation_ground.

[5] https://wiki.openstreetmap.org/wiki/Tag:landuse%3Drecreation_ground.

data by the augmentation with the help of a wordlist. For example, an automatically generated natural language question based on this wordlist asks for bars, whereas the gold parse associated to that question asks for pubs instead:

- Where Bars in Bradford

```
→ query(area(keyval('name','Bradford')),
  nwr(keyval('amenity','pub')),
  qtype(latlong))
```

Conceptually, bars and pubs may not be that different to each other, but OSM advises a strict distinction between bars and pubs [6]. While a pub sells alcohol on premise, a pub also sells food, the athmosphere is more relaxed and the music is quieter compared to a bar.

Lastly, ambiguity was introduced because natural language words now map to multiple different OSM tags. This leads to the following data occurrences:

- shop Off Licenses in Birmingham

```
→ query(area(keyval('name','Birmingham')),
  nwr(keyval('shop','alcohol')),
  qtype(findkey('shop')))
```

- How many closest Off License from Wall Street in Glasgow

```
→ query(around(center(area(
  keyval('name','Glasgow')),
  nwr(keyval('name','Wall Street'))),
  search(nwr(keyval('shop','wine'))),
  maxdist(DIST_
  INTOWN),
  topx(1)),qtype(count))
```

The previous examples show that for the same keyword "Off License" both *shop=alcohol* and *shop=wine* are valid interpretations.

Finally, since the data was augmented first, and only afterwards split into train, development and test sets, there is a lot of overlap between the train and test data. This is problematic because a proper evaluation should also test for overfitting, which does not work if data is shared between different splits, as shown in the following examples:

- Train: cinema in Nantes
- Dev: cinema in Paris
- Test: cinemas in Paris

We applied a dataset de-duplication by removing all datapoints from the development and test sets which are identical to training datapoints when location (e.g., *Paris*) and POI (e.g., *cinema*) are masked. This results in the dataset described in table 3.

System	F1
Lawrence (2018)	80.36
Lawrence (2018)+NER	90.09
token-based	83.43
character-based	93.77

Table 1: F1 results of single-source models on the original NLmaps v2 dataset.

4.2 Semantic Parsing

We use the Joey NMT (Kreutzer et al., 2019) as framework to build a baseline parser. The basic Joey NMT architecture is modified to allow for a multi-source setup (see Figure 3 in the appendix) and for learning from markings.[7]

As evaluation metrics we use exact match accuracy, defined as $\frac{1}{N}\sum_{n=1}^{N}\delta(\text{predicted, gold})$ of a predicted parse and the gold parse. Furthermore, we report F1 score as harmonic mean of recall, defined as the percentage of fully correct answers divided by the set size, and precision, defined as the percentage of correct answers out of the set of answers with non-empty strings.

A character-based Joey NMT semantic parser is able to improve the results reported in Lawrence and Riezler (2018) on the dataset without deduplication, as shown in Table 1. All results presented in the following are relative improvements over our own baseline parser, reported on the deduplicated dataset for which no external baseline is available.

5 Generation of Clarification Questions

On of the goals of error-aware interactive semantic parsing is to alert to user about suspected sources of ambiguity and error by initiating a dialogue. The parser thus needs to detect uncertainty in its output, and generate a clarification questions on the detected source of uncertainty. We use entropy-based uncertainty measures. Firstly, entropy per timestep t is measured as $-\sum_{\tilde{y}_t} p(\tilde{y}_t|x, y_{<t}) \log p(\tilde{y}_t|x, y_{<t})$. This is employed to calculate the entropy of a token as the mean of the character entropies for each of a

[6] https://wiki.openstreetmap.org/wiki/Tag:amenity%3Dbar.

[7] Meta-parameter settings are reported in the appendix.

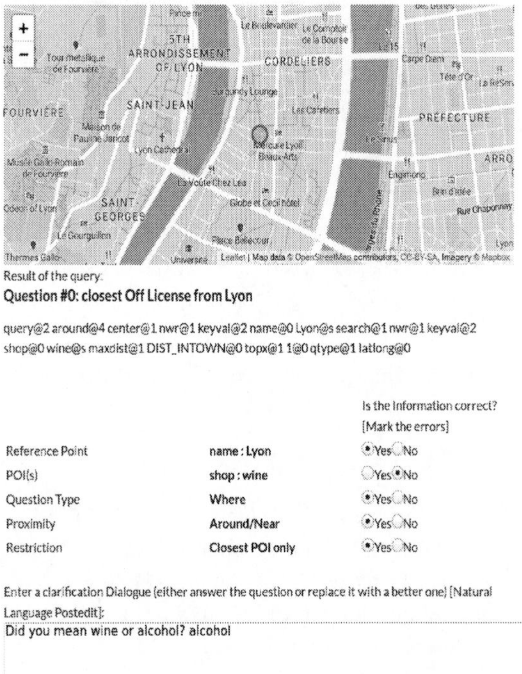

Figure 1: Annotation setup for human interaction study.

System	Accuracy	F1
baseline	83.50	90.26
baseline + hyps	83.66	90.85
baseline + dia	84.74	92.02
baseline + hyps + dia	84.84	91.47
baseline + hyps + dia + log	85.01	91.61

Table 2: Results of the multi-source models compared to the single-source model taking only the source into account on the modified test data.

token's characters.[8] Based on entropy information, we generate simple questions by employing a template-based method which incorporates the least certain token: "Did you mean $token?". Furthermore, we offer alternative answers for the user based on beam search of size 2. This heuristic is justified experimentally since always taking the first beam yields an accuracy of 92.7%, while another 5% of accuracy can be gained by choosing the second beam. This verifies the usefulness of proposing entries in the second beam as alternative in clarification questions: "Did you mean $token or $alternative?".

[8] A visualization of entropy is reported in the appendix.

Split	Count
Train	15,658
Dev	961
Test	4156

Table 3: Statistics of dialogue-enriched data.

6 Experiments on Synthetic Dialogues

In a first experiment, we generated entire dialogues synthetically, that is, the clarification question from the parser and synthetic user answers. The latter were constructed by checking if either the original token or the alternative is contained in the given gold parse. Dataset statistics for train, development and test splits are given in Table 3.

Model training is performed by extending the character-based baseline model by additional encoders for the dialogue (question and answer) and the predicted parse hypothesis. Experiments show that the character-based multi-source model including hypothesis and dialogue as additional input (line 4) outperforms the baseline (line 1) by more than 1 point in accuracy and F1 score (Table 2). This difference is statistically significant with a p-value of 0.0483 determined by approximate randomization.

7 Human Interaction Study

We furthermore performed a small field study where human users interacted with the system. Parses for queries from both train and development parts of the dataset were generated and augmented with automatically created clarification questions based on the uncertainty model. Examples were then filtered to keep only those parses that contained a parse mistake or parse ambiguity. This resulted in a total of 930 annotation tasks.[9]

The annotation interface shown in Figure 1 illustrates the system-user interaction: Human annotators are presented with a natural language query ("closest Off License from Lyon"), the parse (shown below in linearized form), and the result of the generated parse (show as the map extract on top of the figure). In addition to the linearized form of the predicted parse, a human-intelligible list format of the key-value pairs in the parse[10] is presented, following the annotation interface of Lawrence and

[9] Both synthetic and user data will be publicly released.

[10] `https://wiki.openstreetmap.org/wiki/Map_Features`

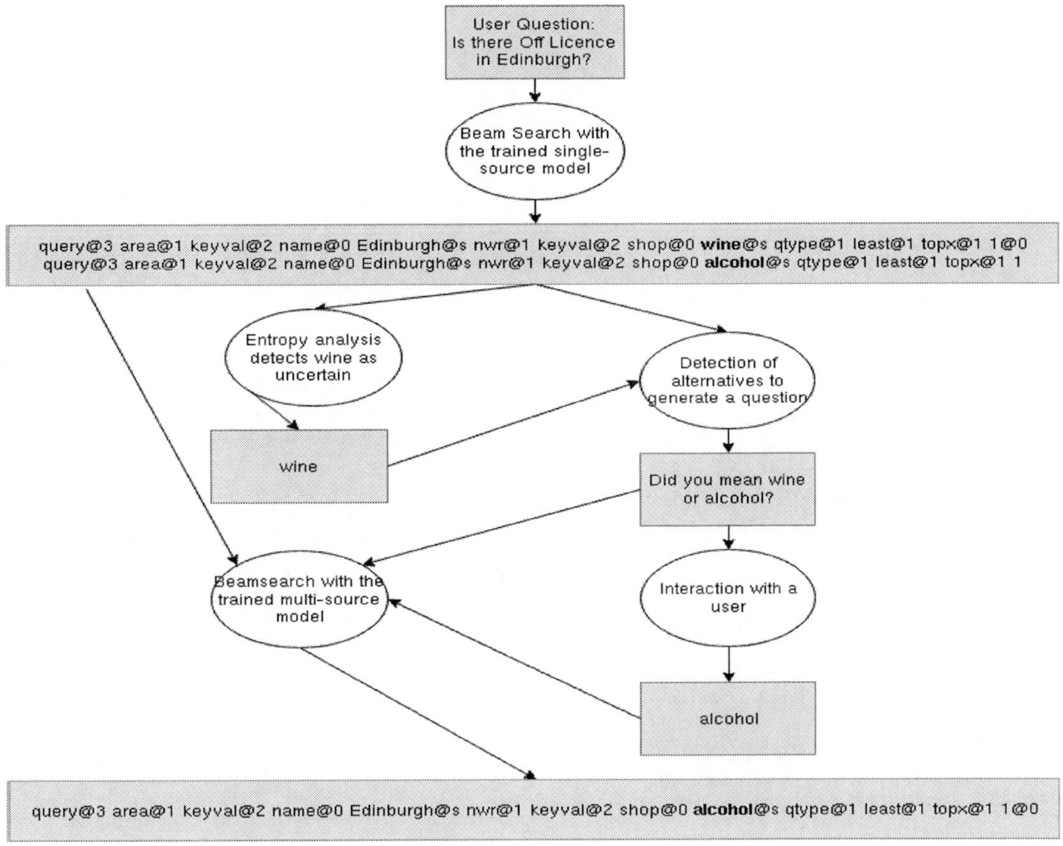

Figure 2: Workflow for the interaction process.

Riezler (2018). The task of the human users is to mark the errors in the list of keys and values, and to answer or correct the clarification question. The markings are used as feedback in the weighted fine-tuning objective of Petrushkov et al. (2018). As the outputs of the model are on character-level, the token-level reward of the annotations is distributed onto them for training. The final model is trained on the weighted objective in a multi-source fashion, taking parse hypothesis, clarification question, and logged user answer as additional inputs. Line 5 in Table 2 shows that fine-tuning a multi-source model that takes hypothesis, dialogue, and logged answer as additional input increases the sequence accuracy by another 0.15%. This difference is statistically significant with a p-value of 0.0027 determined by approximate randomization. The interaction process can be seen in Figure 2.[11]

[11] Additional experiments using the human annotations as test data are reported in the appendix.

8 Conclusion

Ambiguities or errors in real-world semantic OSM parsing arise because of different tagging preferences of developers and users, an issue that can only be solved by an interactive setup where a parser is aware of its errors, and a satisfactory answer is found by the user marking parse errors and communicating alternatives. Our current work is a first step towards precise communication and offline learning in interactive semantic parsing. An interesting future direction of work is to move to online learning in interactive semantic parsing.

Acknowledgments

We would like to thank Christian Buck and Massimiliano Ciaramita for initial fruitful discussions about this work. We would like to thank Christian Buck and Massimiliano Ciaramita for initial fruitful discussions about this work. The research reported in this paper was supported by a Google Focused Research Award on "Learning to Negotiate Answers in Multi-Pass Semantic Parsing".

References

Dzmitry Bahdanau, Kyunghyun Cho, and Yoshua Bengio. 2015. Neural Machine Translation by Jointly Learning to Align and Translate. In *Proceedings of the 3rd International Conference on Learning Representations*.

Ahmed Elgohary, Saghar Hosseini, and Ahmed Hassan Awadallah. 2020. Speak to your Parser: Interactive Text-to-SQL with Natural Language Feedback. In *Proceedings of the 58th Annual Meeting of the Association for Computational Linguistics*, pages 2065–2077.

Julia Kreutzer, Jasmijn Bastings, and Stefan Riezler. 2019. Joey NMT: A Minimalist NMT Toolkit for Novices. In *Proceedings of the 2019 Conference on Empirical Methods in Natural Language Processing and the 9th International Joint Conference on Natural Language Processing: System Demonstrations*, pages 109–114.

Igor Labutov, Bishan Yang, and Tom Mitchell. 2018. Learning to Learn Semantic Parsers from Natural Language Supervision. In *Proceedings of the Conference on Empirical Methods in Natural Language Processing*, pages 1676–1690.

Carolin Lawrence. 2018. *Response-Based and Counterfactual Learning for Sequence-to-Sequence Tasks in NLP*. Universitätsbibliothek Heidelberg, Heidelberg, Germany.

Carolin Lawrence and Stefan Riezler. 2018. Improving a Neural Semantic Parser by Counterfactual Learning from Human Bandit Feedback. In *Proceedings of the 56th Annual Meeting of the Association for Computational Linguistics*, volume 1, pages 1820–1830.

Pavel Petrushkov, Shahram Khadivi, and Evgeny Matusov. 2018. Learning from chunk-based feedback in neural machine translation. In *Proceedings of the 56th Annual Meeting of the Association for Computational Linguistics (ACL)*, Melbourne, Australia.

Ilya Sutskever, Oriol Vinyals, and Quoc Le. 2014. Sequence to Sequence Learning with Neural Networks. In Z. Ghahramani, M. Welling, C. Cortes, N. D. Lawrence, and K. Q. Weinberger, editors, *Advances in Neural Information Processing Systems*, volume 27, pages 3104–3112. Curran Associates, Inc., Montreal, Canada.

Ziyu Yao, Xiujun Li, Jianfeng Gao, Brian M. Sadler, and Huan Sun. 2019. Interactive Semantic Parsing for If-Then Recipes via Hierarchical Reinforcement Learning. In *Proceedings of the 33rd Conference on Artificial Intelligence (AAAI)*, pages 2547–2554.

Tao Yu, Rui Zhang, He Yang Er, Suyi Li, Eric Xue, Bo Pang, Xi Victoria Lin, Yi Chern Tan, Tianze Shi, Zihan Li, Youxuan Jiang, Michihiro Yasunaga, Sungrok Shim, Tao Chen, Alexander Fabbri, Zifan Li, Luyao Chen, Yuwen Zhang, Shreya Dixit, Vincent Zhang, Caiming Xiong, Richard Socher, Walter Lasecki, and Dragomir Radev. 2019a. CoSQL: A Conversational Text-to-SQL Challenge Towards Cross-Domain Natural Language Interfaces to Databases. In *Proceedings of the 2019 Conference on Empirical Methods in Natural Language Processing and 9th International Joint Conference on Natural Language Processing*, pages 1962–1979.

Tao Yu, Rui Zhang, Kai Yang, Michihiro Yasunaga, Dongxu Wang, Zifan Li, James Ma, Irene Li, Qingning Yao, Shanelle Roman, Zilin Zhang, and Dragomir Radev. 2018. Spider: A Large-Scale Human-Labeled Dataset for Complex and Cross-Domain Semantic Parsing and Text-to-SQL Task. In *Proceedings of the 2018 Conference on Empirical Methods in Natural Language Processing*, pages 3911–3921.

Tao Yu, Rui Zhang, Michihiro Yasunaga, Yi Tan, Victoria Lin, Suyi Li, Heyang Er, Irene Li, Bo Pang, Tao Chen, Emily Ji, Shreya Dixit, David Proctor, Sungrok Shim, Jonathan Kraft, Vincent Zhang, Caiming Xiong, Richard Socher, and Dragomir Radev. 2019b. SParC: Cross-Domain Semantic Parsing in Context. In *Proceedings of the 57th Annual Meeting of the Association for Computational Linguistics*, pages 4511–4523.

Barret Zoph and Kevin Knight. 2016. Multi-Source Neural Translation. In *Proceedings of the 2016 Conference of the North American Chapter of the Association for Computational Linguistics: Human Language Technologies*, pages 30–34.

A Supplementary Material for "Towards Error-Aware Interactive Semantic Parsing"

A.1 Hyperparameter Settings

A.2 Evaluation on the human annotated data

In an additional experiment, we evaluated the models that were trained on the synthetically generated dataset on the data resulting from the human interaction study. The result of comparing the baseline model with the multi-source model trained on parse hypothesis and synthetic dialogue as additional inputs is shown in Table 5. The astonishing gains of over 15% in F1 score can be explained by the fact that the data for human annotation set were filtered to include only examples for which the baseline parser did not match the gold standard parse (thus producing an accuracy score of 0).

System	Accuracy	F1
char	0	43.07
char+ hyps + dia	25.09	60.85

Table 5: Test results on human-annotated data.

A.3 Entropy visualization

The entropy of the parse of the sentence "How many Off License in Heidelberg" can be seen in Figure 4. The character-based model shows uncertainty with respect to the token *wine*. This is the desired result because the alternative for this position would be *alcohol*.

Parameter	Lawrence and Riezler (2018)	token-based	character-based
Attention mechanism	bahdanau	bahdanau	bahdanau
RNN type	gru	gru	gru
Embedding size	1000	620	620
Encoder layer count	1	1	1
Encoder hidden size	1024	400	400
Decoder layer count	1	1	1
Decoder hidden size	1024	800	800

Table 4: Parameter overview compared to Lawrence and Riezler (2018).

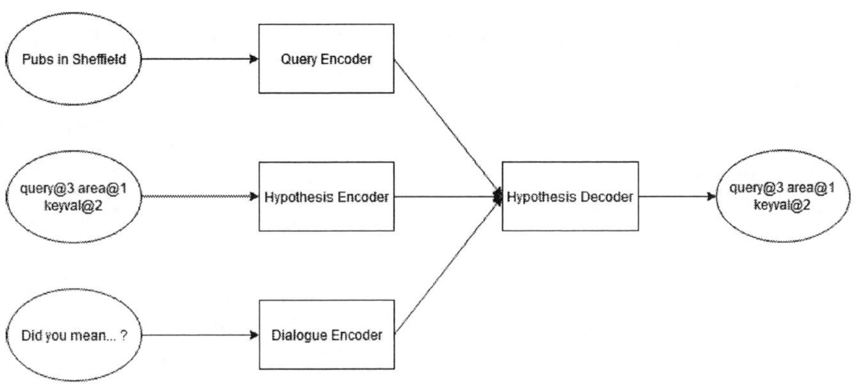

Figure 3: Multi-source semantic parsing.

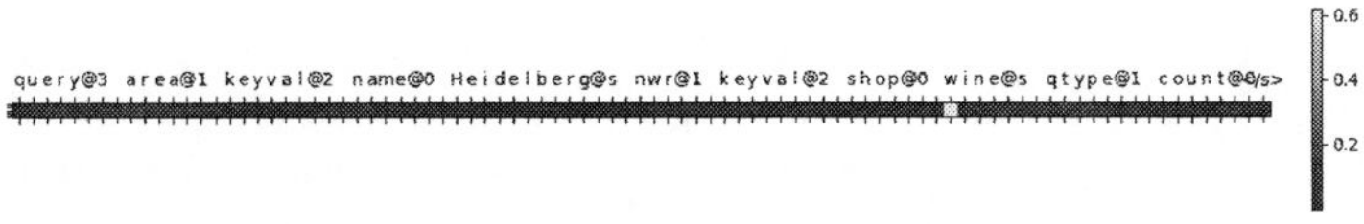

Figure 4: Character entropy for parse of query "How many Off License in Heidelberg".

Plan Explanations that Exploit a Cognitive Spatial Model

Raj Korpan
The Graduate Center
City University of New York
rkorpan@gradcenter.cuny.edu

Susan L. Epstein
The Graduate Center and Hunter College
City University of New York
susan.epstein@hunter.cuny.edu

Abstract

Ideally, people who navigate together in a complex indoor space share a mental model that facilitates explanation. This paper reports on a robot control system whose cognitive world model is based on spatial affordances that generalize over its perceptual data. Given a target, the control system formulates multiple plans, each with a model-relevant metric, and selects among them. As a result, it can provide readily understandable natural language about the robot's intentions and confidence, and generate diverse, contrastive explanations that reference the acquired spatial model. Empirical results in large, complex environments demonstrate the robot's ability to provide human-friendly explanations in natural language.

1 Introduction

Inspired by recent recommendations for spoken language interaction with robots (Marge et al., 2020), this paper introduces WHY, an approach to communicate a robot's planning rationales, intentions, and confidence in human-friendly spatial language. Our thesis is that a plan based on spatial representations acquired from travel experience can ground its objectives and support explainable path planning. The principal results of this paper are empirical demonstrations of WHY's ability to explain and contrast plans in readily-understandable natural language.

Given sensor data and a metric map (e.g., a floor plan), the task of our autonomous robot navigator is to travel to target locations in a large, complex, human-centric, indoor space (henceforward, *world*). The robot's control system integrates acquired spatial knowledge into a cognitively-based architecture that combines planning with reactivity, heuristics, and situational reasoning. Given a target, the control system creates a *plan*, a sequence of intermediate locations (*waypoints*) to reach it. This plan is expected to balance multiple objectives, combine continuous and discrete spatial representations, and encourage a human's trust.

Traditional navigation planners use a *cost graph* (also known as a *costmap*) where each node is a point in unobstructed space and each edge connects a pair of nodes with a weight for the cost to move between them. A popular cost graph is based on an *occupancy grid*, uniform square cells superimposed on a two-dimensional metric map. Each edge in the graph represents two adjacent unobstructed cells, labeled with the Euclidean distance between their centers. In a fine-grained grid, however, optimal planners (e.g., $A*$ (Hart et al., 1968)) hug obstacles so tightly that their plans require tight maneuvers to reach some waypoints and may fail as actuator and sensor errors accumulate near them.

To bias plans toward its particular *objective* (a spatial representation or commonsense rationale), a planner modifies the weights in its own copy of the occupancy-grid graph. The fixed underlying graph structure allows our approach to evaluate a plan within any such modified graph. Voting then selects the plan that best satisfies all the objectives. This approach facilitates contrastive natural-language explanations of the chosen plan with respect to each objective. The control system reports on its beliefs, intentions, and confidence with spatial language. For example, "Although there may be another way that is somewhat shorter, I think my way is a lot better at going through open areas."

The next sections provide related work and describe the acquired spatial model. Subsequent sections cover the modified graphs, vote-based planning, and how WHY explains plans. The last sections describe empirical results and future work.

2 Related work

A spatial representation of its world is essential to a robot control system that navigates efficiently and explains its behavior clearly. Grounded communication between a robot and a person, however, requires a shared spatial representation. This section first describes work on human cognitive maps

Proceedings of Second International Combined Workshop on Spatial Language Understanding andGrounded Communication for Robotics, pages 60–70
August 5–6, 2020.

that inspired our control system's spatial model. It then details approaches that describe and explain the robot's behavior.

A *cognitive map* is a compact, mental spatial representation of a world, built by a person as she moves through that world (Golledge, 1999). To reduce her cognitive load, a person reasons from a cognitive map that incorporates landmarks, route knowledge, and survey knowledge (Tversky, 1993). Landmarks represent locations in the map, routes represent lines that connect them, and survey knowledge captures spatial relations. Although it has been suggested that cognitive maps use metric distances and angles (Gallistel, 1990), more recent work indicates that cognitive maps have a non-metric, qualitative topological structure (Foo et al., 2005). Other recent work suggests that people use a cognitive graph with labeled metric information that captures connectivity and patterns (Chrastil and Warren, 2014; Warren et al., 2017).

An *affordance* is a characteristic of the world that enables the execution of some action (Gibson, 1977). Affordance-based theories of spatial cognition posit a tight relationship between the specific dynamics of a world and the decisions made by an individual there (Fajen and Phillips, 2013). Here, a *spatial affordance* is an abstract representation of the world that facilitates navigation. This paper introduces path planning in cost graphs based on acquired spatial affordances. People generalize structured representations across domains on similar tasks (Pouncy et al., 2021) much the way the spatial model described here generalizes affordances for use in different worlds.

A control system can learn and use a cognitive map of its world for robot navigation. For example, the Spatial Semantic Hierarchy (SSH) modeled a cognitive map with hierarchical metric and topological representations (Kuipers, 2000). Although SSH's cognitive map bears some similarity to the one used here, it did not explain plans. Other approaches used semantics to create a meaningfully-labeled metric map (Kostavelis and Gasteratos, 2015). While these maps provide a qualitative context in which to ground a controller's language, they do not necessarily align with human cognitive maps. Moreover, control systems often use semantic maps for communication but another representation for reasoning and decision-making. Instead, this paper shows how a single, affordance-based representation supports all of those processes.

Indoors, an autonomous robot may interact with people as it navigates to its target. A human collaborator is more likely to accept, trust, and understand a robot that can explain its behavior (Rosenfeld and Richardson, 2019). Rather than describe an event or summarize its causes, an explanation compares counterfactual cases, includes causes selectively, and recognizes people as social beings with beliefs and intentions (Miller, 2019). A *contrastive* explanation compares the reason for a decision to another plausible rationale (Hoffmann and Magazzeni, 2019). Human subjects generally prefer such explanations that focus on the difference between the robot's planned route and their own (e.g., "my route is shorter, but overlaps more and produces less reward") (Perelman et al., 2020).

Detailed technical logs of a robot's experience were originally available only to trained researchers (Landsiedel et al., 2017; Scalise et al., 2017). Recent work, however, has generated natural language descriptions of a robot's travelled path from them. These focus on abstraction, specificity, and locality (Rosenthal et al., 2016; Perera et al., 2016) or on sentence correctness, completeness, and conciseness (Barrett et al., 2017). All, however, required a labeled dataset or a semantic map. Other recent work partitions a plan into actions and uses language templates to generate descriptions of each action in the context of a collaborating robot team (Singh et al., 2021). WHY focuses on explanations for the reasons behind the robot's decisions rather than descriptions of the robot's behavior.

To produce explanations, others have selected potentially suboptimal plans (Fox et al., 2017; Chakraborti et al., 2019) or readily understandable behaviors (Huang et al., 2019), or relied on classical planning (Magnaguagno et al., 2017; Grea et al., 2018; Krarup et al., 2019) or on logic (Seegebarth et al., 2012; Nguyen et al., 2020). None of that work, however, explains in natural language. The approach closest to the one presented here provides contrastive explanations for multi-objective path planning in natural language as a Markov decision process (Sukkerd et al., 2020), but considers fewer objectives, requires a hand-labeled map, and has been evaluated only in much smaller worlds.

3 Spatial affordances

The context of this work is *SemaFORR*, a cognitively-based control system for autonomous indoor navigation (Epstein et al., 2015; Epstein and

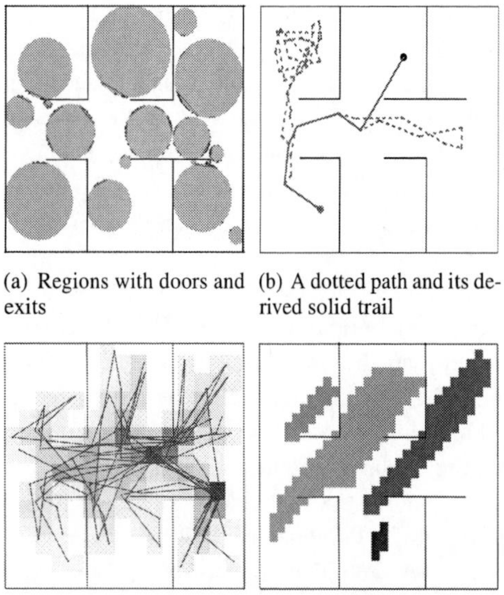

(a) Regions with doors and exits (b) A dotted path and its derived solid trail

(c) Trails and conveyors (d) Diagonal hallways

Figure 1: Affordances in a simple artificial world

Korpan, 2019). At *decision point* $d = \langle x, y, \theta, V \rangle$, SemaFORR records the robot's location (x, y), its orientation θ, and its *view* V, the data from its onboard range finder. After each target, SemaFORR identifies spatial affordances for its acquired model of *freespace*, the unobstructed areas in a world. The model can be used alone or with a metric map.

At decision point d, SemaFORR learns a *region*, a circle in freespace with center at (x, y) and radius equal to the minimum distance reported by V. Accumulated contradictory or overlapping regions are resolved after each target. An *exit* represents access to freespace, a point where the robot's path once crossed the region's perimeter. A *door* is an arc on a region's perimeter, a continuous generalization of finitely many, relatively close exits between its endpoints. Figure 1(a) shows acquired regions with exits and doors (drawn for clarity as secants to their respective arcs). Although regions approximate what appear to be rooms in the figure, they record only freespace, not walls.

A *trail* is a refined version of the robot's path toward its target. The algorithm that creates trails heuristically smooths the robot's paths and eliminates digressions. The remaining (usually far fewer) decision points are *trail markers*. As in Figure 1(b), the sequence of line segments defined by consecutive trail markers is typically more direct than the original path, but rarely optimal. A *conveyor* is a freespace cell in a $2 \times 2m$ grid superimposed on the world's footprint. Conveyors tally how often trails pass through them. Higher-count cells represent locations that frequently support travel. They appear darker in Figure 1(c).

Table 1: SemaFORR's planners and their objectives

Planner	Objective
FASTP	Minimize distance traveled
SAFEP	Avoid obstacles
EXPLOREP	Avoid paths
NOVELP	Avoid spatial model
CONVEYP	Exploit conveyors
HALLWAYP	Exploit hallways
REGIONP	Exploit regions, doors, exits
TRAILP	Exploit trail markers

A hallway represents well-travelled routes in some *angular direction* (vertical, horizontal, major diagonal, or minor diagonal). A hallway generalizes line segments between consecutive decision points to find relatively straight, narrow, continuous freespace with both length and width. Figure 1(d) shows some acquired minor-diagonal hallways.

4 Modified cost graphs

Planning for navigation requires a graphical representation of the world's freespace. To produces an optimal plan, A* searches a cost graph G based on an occupancy grid with edge weights for Euclidean distance. SemaFORR constructs a set of graphs; each begins with G but modifies its edge weights to align with a particular objective. This biases search toward that objective but still considers plan length. In practice, an occupancy grid should be sufficiently fine to represent obstacles accurately.

Table 1 lists SemaFORR's planners and their objectives. Given a target, each planner formulates its own plan to reach it, one biased toward its own objective. Two planners focus on commonsense: FASTP searches the original G, but SAFEP increases G's edge weights based on an edge's proximity to obstacles. Two others focus on exploration to acquire more knowledge about their world. EXPLOREP creates a grid that tallies how frequently the robot's path history passes through each cell, and uses those values to increase edge weights where it has already traveled. Because the acquired spatial model summarizes experience more compactly than a path, NOVELP explores areas not covered by the model. It increases a weight if the edge overlaps an acquired affordance.

Four planners exploit a particular kind of spatial affordance with changes to edge weights. (Values based on preliminary testing bias plans to pursue but not overemphasize affordances.) REGIONP's cost graph modifies each edge's weight w based on the location of its endpoints. If both lie in the same region, w goes unchanged; if neither lies in a region w becomes $10w$. Otherwise, for the one endpoint v not in a region, w becomes $1.5w$ if v is within $0.5m$ of a door and an exit, $1.75w$ if v is within $0.5m$ of a door or an exit, and otherwise $2w$. This biases plans to pass through regions because it increases edge costs outside them.

HALLWAYP and TRAILP modify their weights similarly, with respective conditions "lie in one hallway" and "lie within $0.5m$ of a trail marker." If both endpoints of an edge meet the condition, w goes unchanged; if neither does, w becomes $10w$. Otherwise, when just one endpoint meets the condition, w becomes $1.5w$. To bias plans toward high-count conveyors, CONVEYP considers the counters c_1 and c_2 for the cells where the endpoints of an edge with weight w lies. If both are non-zero, w becomes $w + 2/(c_1 + c_2)$; otherwise, w becomes $10w$.

Because SemaFORR's spatial model focuses on freespace, these modified cost graphs allow a robot control system to encourage travel there but also incorporate the metric cost graph where the model lacks knowledge. The region-based cost graph, for example, imposes relatively lower costs only for doors and exits that the robot has successfully exploited earlier, and thus prioritizes them. Because weights only increase, Euclidean distance remains an *admissible* heuristic for A*, that is, it never overestimates the actual cost to the target's location.

5 Voting among planners

To choose paths, people use many different objectives that reflect their motivation (Golledge, 1999). A cognitively-based robot navigator should also incorporate and balance a variety of path-selection heuristics. SemaFORR's planners can be used together because they originate from the same cost graph. This section explains Algorithm 1, pseudocode for how voting balances the planners' objectives to select a plan.

SemaFORR constructs multiple plans that optimize a single objective and then uses voting to select the plan that maximally satisfies the most objectives. First, each planner j constructs an optimal plan P_j for its objective as a sequence of waypoints in its modified cost graph G_j. This guarantees that each submitted plan is optimal for at least one objective.

Algorithm 1: Voting-based planning

Input: *planners J, spatial model M, basic cost graph G*
for *each planner $j \in J$* **do**
    Set j's cost graph G_j to a copy of G
    Modify G_j's weights based on j and M
    With A*, find optimal plan P_j in G_j
for *each planner $j \in J$* **do**
    for *each planner $i \in J$* **do**
        $C_{ij} \leftarrow$ cost of plan P_i in G_j
    Normalize plan scores C_{ij} in [0,10]
for *each plan P_i* **do**
    $Score_i \leftarrow \sum_{j=1}^{J} C_{ij}$
$best \leftarrow argmin_i \; Score_i$
return P_{best}

Next, each planner's objective is used to evaluate every plan. All the cost graphs have the same nodes and edges, so to evaluate planner i's plan P_i from the perspective of planner j, SemaFORR simply sums the edge weights in G_j for the sequence of edges specified by P_i. The resultant scores C_{ij} are then normalized in $[0, 10]$ for each j. SemaFORR seeks to minimize its objectives. Thus a C_{ij} value near 0 indicates that plan P_i closely conforms to objective j, while a score near 10 indicates that plan P_i conflicts with objective j. Voting selects the plan with the lowest total score across all objectives and breaks ties at random.

6 Contrastive explanations

SemaFORR uses WHY to explain its long-range perspective in natural language. WHY exploits differences among planners' objectives to produce clear, concise, contrastive explanations for a plan quickly. WHY assumes that the robot's human companion seeks a shortest-length plan, and compares that to SemaFORR's plan. Although we assume here that a goal-directed human navigator would seek to minimize travel distance, another objective, including those in Table 1, could label the foundational cost graph G instead.

Throughout this section, $\mathcal{N}$ represents a function that translates its argument (a planner or a metric value) into natural language. Given a real-valued

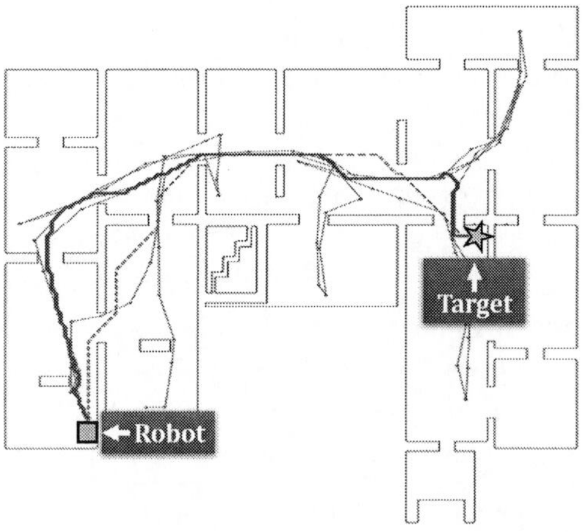

Figure 2: WHY compares FASTP's (red) plan to TRAILP's (blue) one biased by SemaFORR's (green) trails. It explains, "Although there may be another way that is a lot shorter, I think my way is a lot better at following ways we've gone before."

metric m for some aspect (e.g., confidence or enthusiasm) of the decision process, $\mathcal{M}(m)$ bins m's value into an ordered partition of m's range and $\mathcal{N}(\mathcal{M}(m))$ translates that bin to a natural language phrase. For example, m could measure the desire to select one plan over the others, and the value partition could distinguish a strong preference for that plan from a weak one. Thus, if $m \in (0, +\infty)$ were partitioned as $\{(0, 5), [5, +\infty)\}$, $\mathcal{N}(m < 5)$ could be "a little" and $\mathcal{N}(m \geq 5)$ "a lot." This allows WHY to hedge in its responses, much the way people explain their reasoning when they are uncertain (Markkanen and Schröder, 1997).

6.1 Why does your plan go this way?

Human and robot plans to reach the same target may differ because they lack a common objective. WHY's response to this question presumes that a human plans from one perspective, objective β_H, while the robot plans from another perspective, objective β_R. Explanations for a plan assume a human has an alternative objective. Henceforward, β_H is "take the shortest path."

WHY models the human questioner with β_H to produce plan P_H, a prediction of the human's implicit plan. Algorithm 2 is pseudocode for WHY's plan-explanation procedure. WHY takes as input the robot's plan P_R and objective β_R, and the alternative plan P_H and objective β_H it attributes to the human questioner. $\beta_H(P)$ measures plan length and $\beta_R(P)$ measures plan cost in P_R's graph. In the running example shown in Figure 2, WHY ex-

Algorithm 2: Explanation procedure

Input: *planning objectives* β_R *and* β_H, *plans* P_R *and* P_H
Output: *explanation*
$\mathcal{D}_R = \beta_R(P_R) - \beta_R(P_H)$
$\mathcal{D}_H = \beta_H(P_R) - \beta_H(P_H)$
switch *mode*($\mathcal{D}_R$, $\mathcal{D}_H$) **do**
  case $\mathcal{D}_R = \mathcal{D}_H = 0$ **do**
    explanation $\leftarrow$ sentence based on template for equivalent plans
  case $\mathcal{D}_R < 0$ *and* $\mathcal{D}_H > 0$ **do**
    explanation $\leftarrow$ sentence for β_R, β_H
  case $\mathcal{D}_R < 0$ *and* $\mathcal{D}_H = 0$ **do**
    explanation $\leftarrow$ sentence for β_R
return *explanation*

plains SemaFORR's preference for its plan P_R from TRAILP where β_R is TRAILP's objective ("exploit trail markers"). WHY translates β_H and β_R with Table 2 as "short" and "follows ways we've gone before," respectively.

If voting selected the plan constructed by FASTP (i.e., the shortest-length plan), then Why responds with "I decided to go this way because I agree that we should take the shortest route." Otherwise, to compare P_R with P_H, WHY calculates their difference from two perspectives: $\mathcal{D}_H$ from the human's perspective (e.g., length), and $\mathcal{D}_R$ from the robot's perspective (e.g., proximity to trails). WHY places these differences in user-specified bins that represent a human perspective on the objectives. Table 3 provides language for these differences.

The relative size of the differences determines an applicable template. If both $\mathcal{D}_H$ and $\mathcal{D}_R$, as defined in Algorithm 2, are 0, then the plans equally address the two objectives, and WHY explains "I decided to go this way because I think it's just as $\mathcal{N}(\beta_H)$ and equally $\mathcal{N}(\beta_R)$." Otherwise, the plans differ with respect to one or both objectives. If $\mathcal{D}_R$ is negative (e.g., P_R is more aligned with trails), then WHY instantiates this template:

1: Although there may be another way that is $\mathcal{N}(\mathcal{M}(\mathcal{D}_H))\ \mathcal{N}^*(\beta_H)$,
2: I think my way is $\mathcal{N}(\mathcal{M}(\mathcal{D}_R))\ \mathcal{N}^*(\beta_R)$.

where $\mathcal{N}^*(\beta)$ is a comparator for β (e.g., "shorter" or "better at following ways we've gone before"). For example, "Although there may be another way that is somewhat shorter, I think my way is a lot better at following ways we've gone before." WHY omits line 1 in the template if $\mathcal{D}_H = 0$. Other cases,

Table 2: Language for the planners' objectives. $\mathcal{N}^*(\beta)$ and $\mathcal{N}'(\beta)$ values for FASTP and EXPLOREP are as shown. For the others, $\mathcal{N}^*(\beta) \approx \mathcal{N}'(\beta)$, where $\mathcal{N}^*(\beta)$ begins with "better at" and $\mathcal{N}'(\beta)$ begins with "worse at."

Planner	$\mathcal{N}(\beta)$	$\mathcal{N}^*(\beta)$	$\mathcal{N}'(\beta)$
FASTP	short	shorter	longer
EXPLOREP	goes a new way	newer	familiar
SAFEP	stays far from obstacles	staying far from obstacles	
NOVELP	learns something new	learning something new	
CONVEYP	goes through well-traveled areas	going through well-traveled areas	
HALLWAYP	follows hallways	following hallways	
REGIONP	goes through open areas	going through open areas	
TRAILP	follows ways we've gone before	following ways we've gone before	

Table 3: Language for value intervals for the difference $\mathcal{D}$. For affordance-based planners a=150 and b=25, for SAFEP a=0.35 and b=0.15, for EXPLOREP a=100 and b=15, and for NOVELP a=350 and b=100.

Planner	Intervals $\mathcal{M}(\mathcal{D})$	$\mathcal{N}(\mathcal{M}(\mathcal{D}))$
	$(0, 1]$	a bit
FASTP	$(1, 10]$	somewhat
	$(10, +\infty)$	a lot
	$(-\infty, -a]$	a lot
All others	$(-a, -b]$	somewhat
	$(-b, +\infty)$	a bit

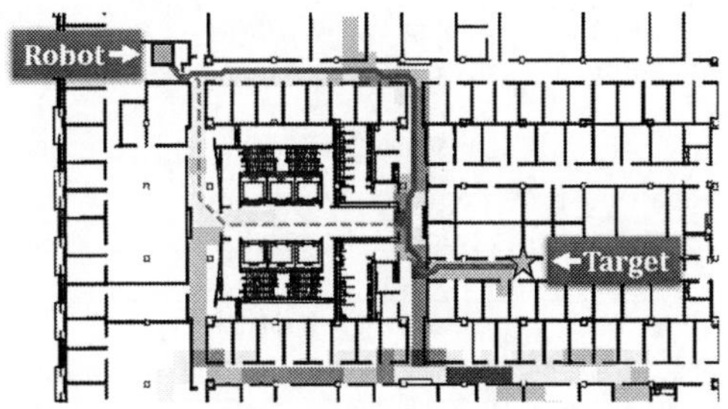

Figure 3: Acquired conveyors in green, with darker higher-count cells. Voting chose CONVEYP's (blue) plan which is drawn to high-count cells. In response to "What's another way we could go?" WHY compares the conveyor plan with FASTP's (red) plan: "We could go that way since it's a bit shorter but it could also be a bit worse at going through well-traveled areas."

where $\mathcal{D}_H < 0$ or $\mathcal{D}_R > 0$ cannot occur because each planner is optimal with respect to its own cost graph and objective, as described in Section 5.

6.2 Why do you prefer your plan?

WHY also addresses the question "Why do you prefer your plan?" Unlike the previous response, which contrasted the human's objective with the robot's, this response has the robot explain its objective. If voting selects the FASTP plan, which the robot assumes has the same objective as its human companion, WHY would respond "Actually, I agree that we should take the shortest route." Otherwise, WHY uses the differences $\mathcal{D}_H$ and $\mathcal{D}_R$ from Algorithm 2. If they are both 0, then WHY replies, "I think both plans are equally good." Otherwise, WHY responds with the template "I prefer my plan because it's $\mathcal{N}(\mathcal{M}(\mathcal{D}_R))$ $\mathcal{N}^*(\beta_R)$." For example, to explain why SemaFORR chose TRAILP's plan, WHY might say "I prefer my plan because it's a lot better at following ways we've gone before."

6.3 What's another way we could go?

Figure 3 shows an example where WHY responds to "What's another way we could go?" Because WHY has access to two plans from SemaFORR (P_R and P_H), it can provide P_H, the shortest-path plan, as the alternative plan in response. If voting selects the FASTP plan, which uses the same objective as the robot's human companion, then WHY responds "Yours is the best way to go." Otherwise, it instantiates the template: "We could go your way since it's $\mathcal{N}(\mathcal{M}(\mathcal{D}_H))$ $\mathcal{N}^*(\beta_H)$ but it could also be $\mathcal{N}(\mathcal{M}(\mathcal{D}_R))$ $\mathcal{N}'(\beta_R)$." Here $\mathcal{N}'$ denotes an opposite comparator (e.g., "longer" or "worse at following ways we've gone before"). For example, an explanation is "We could go that way since it's somewhat shorter but it could also be a lot worse at following ways we've gone before."

6.4 How sure are you about your plan?

In response to "How sure are you about your plan?" WHY explains its confidence that P_R meets its objective. Figure 4 shows an example. WHY uses the language for $\mathcal{M}(\mathcal{D}_R)$ and $\mathcal{M}(\mathcal{D}_H)$ from Table 3 to extract a value $\mathcal{C} = \mathcal{N}(\mathcal{M}(\mathcal{D}_R, \mathcal{D}_H))$ from Table 4. WHY then instantiates "I'm $\mathcal{N}(\mathcal{C})$ sure because" followed by line $\mathcal{C}$:

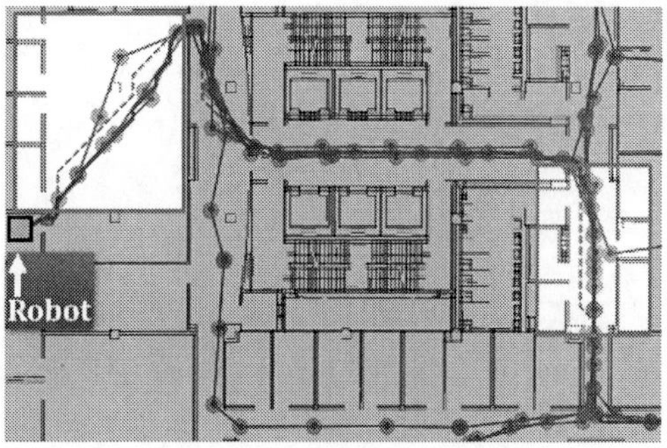

Figure 4: Highlighted sections of FASTP's (red) plan and TRAILP's (blue) plan to follow acquired (green circle) trail markers. WHY explains "I'm really sure because my plan is a lot better at following ways we've gone before and only a bit longer than your plan."

Table 4: Language $\mathcal{N}(\mathcal{M}(\mathcal{D}_R, \mathcal{D}_H))$ for confidence compares $\mathcal{M}(\mathcal{D}_R)$ and $\mathcal{M}(\mathcal{D}_H)$ from Table 3. Here, 1 denotes "really," 2 = "only somewhat," and 3 = "not."

	$\mathcal{N}(\mathcal{M}(\mathcal{D}_H))$		
$\mathcal{N}(\mathcal{M}(\mathcal{D}_R))$	"a lot"	"somewhat"	"a bit"
"a lot"	2	1	1
"somewhat"	3	2	1
"a bit"	3	3	2

1: my plan is $\mathcal{N}(\mathcal{M}(\mathcal{D}_R))$ $\mathcal{N}^*(\beta_R)$ and only $\mathcal{N}(\mathcal{M}(\mathcal{D}_H))$ $\mathcal{N}'(\beta_H)$ than yours.
2: even though my plan is $\mathcal{N}(\mathcal{M}(\mathcal{D}_R))$ $\mathcal{N}^*(\beta_R)$, it is also $\mathcal{N}(\mathcal{M}(\mathcal{D}_H))$ $\mathcal{N}'(\beta_H)$ than yours.
3: my plan is $\mathcal{N}(\mathcal{D}_H)$ $\mathcal{N}'(\beta_H)$ and only $\mathcal{N}(\mathcal{D}_R)$ $\mathcal{N}^*(\beta_R)$ than yours

6.5 How are we getting there?

"How are we getting there?" shows a human companion's uncertainty about the route planned to reach their shared target. Rather than reference the planner's objective, WHY treats this as a request for a high-level description of P_R itself, and uses the segments between consecutive waypoints in SemaFORR's plan P_R to produces natural language that describes it. Figure 5 shows an example.

WHY anticipates travel with P_R as an ordered sequence of locations from the robot's current location through P_R's waypoints and then to the target. First, WHY forms plan segments from consecutive locations in P_R and computes each segment's length and angular direction χ (based on the angle between its endpoints relative to a fixed horizontal axis). It then bins χ within an interval $\mathcal{M}(\chi)$ and assigns a label $\mathcal{N}(\mathcal{M}(\chi))$ as shown in Table 5.

These labels are allocentric, and therefore less

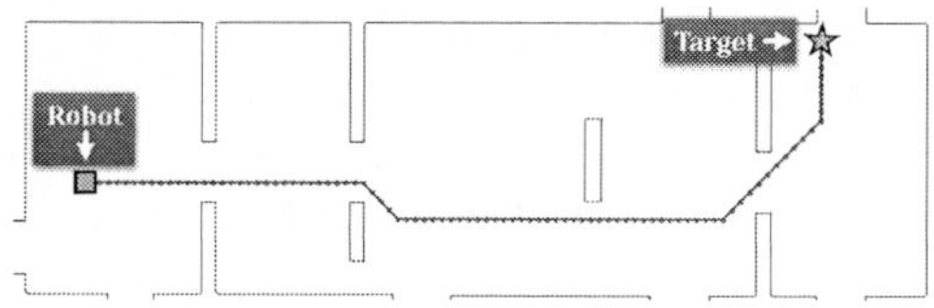

Figure 5: SemaFORR's FASTP plan with 92 waypoints from the robot to its target. WHY explains in 9 clauses, "We will go straight about 20 meters, turn right a little, go straight about 4 meters, turn left a little, go straight about 20 meters, turn left a little, go straight about 8 meters, turn left a little, and go straight about 4 meters to reach our target."

Table 5: Labels $\mathcal{N}(\mathcal{M}(\chi))$ for segment angle intervals $\mathcal{M}(\chi)$. Language $\mathcal{N}(\alpha)$ adjusts the change in consecutive angular directions for full 2π rotation: $\alpha = \mathcal{N}(\mathcal{M}(\chi_k)) - \mathcal{N}(\mathcal{M}(\chi_{k-1}))) \bmod 8$.

$\mathcal{M}(\chi)$	$\mathcal{N}(\mathcal{M}(\chi))$	α	Phrase $\mathcal{N}(\alpha)$
$[\frac{-7\pi}{8}, \frac{-5\pi}{8})$	2	0	go straight
$[\frac{-5\pi}{8}, \frac{-3\pi}{8})$	3	1	turn left a little
$[\frac{-3\pi}{8}, \frac{-\pi}{8})$	4	2	turn left
$[\frac{-\pi}{8}, \frac{\pi}{8})$	5	3	turn hard left
$[\frac{\pi}{8}, \frac{3\pi}{8})$	6	4	turn around
$[\frac{3\pi}{8}, \frac{5\pi}{8})$	7	5	turn hard right
$[\frac{5\pi}{8}, \frac{7\pi}{8})$	8	6	turn right
otherwise	1	7	turn right a little

appropriate indoors. WHY translates them to an egocentric frame of reference, as if the robot and its companion faced the same way along the intended route. The change in consecutive $\mathcal{N}(\mathcal{M}(\chi))$ labels represents the change in direction from one path segment to the next. $\mathcal{N}(\alpha)$ is language for α, the angular change in χ from one segment to the next. For example, if the first segment in P_R were labeled 2 and the second segment labeled 7, then $\alpha = 5$ which Table 5 translates as "turn hard right."

Plan P_R now has a sequence of phrases for the points where two consecutive segments meet. WHY inserts a "go straight" after each "turn" phrase. WHY then summarizes consecutive "go straight" phrases into a single one (since they indicate no change in direction) with a length $\mathcal{L}$, the sum of the lengths of the segments that induced it. These $\mathcal{L}$s are binned into intervals and reported in natural language (e.g., $5.7m$ lies in $(4, 6]$ with language "about 6 meters").

WHY combines the list of phrases and lengths appropriately to form a succinct explanation with the template "We will [$\mathcal{N}(\alpha)$ {about $\mathcal{N}(\mathcal{M}(\mathcal{L}))$},] to reach our target." It repeats the material in square

Table 6: How often planners won the vote

Planner	M5	H10	G5	Total
FASTP	25.0%	42.9%	32.4%	33.4%
SAFEP	37.0%	25.7%	27.5%	30.1%
EXPLOREP	9.0%	6.9%	4.9%	6.9%
NOVELP	0.0%	0.0%	0.0%	0.0%
CONVEYP	14.0%	7.4%	16.5%	12.6%
HALLWAYP	6.0%	9.1%	6.6%	7.2%
REGIONP	5.5%	6.3%	0.5%	4.1%
TRAILP	3.5%	1.7%	11.5%	5.6%

brackets for each $\mathcal{N}(\alpha)$, and includes the material in curly brackets only when $\mathcal{N}(\alpha)$ is "go straight."

In summary, WHY produces natural explanations for a robot's plan as it travels through a complex world. These explanations are essential for human-friendly autonomous indoor navigation and require an assumption about its human collaborator's objective. Our approach explains the robot's plan, responds to questions about alternatives, and expresses a human-friendly level of confidence.

7 Empirical Evaluation

SemaFORR with WHY is evaluated on three challenging real worlds: M5, H10, and G5. M5 is the fifth floor of New York's Museum of Modern Art. It is $54 \times 62m$ and has $1585m^2$ freespace. H10 is the $89 \times 58m$ tenth floor of an academic building with $2627m^2$ of freespace and 75 rooms. G5 is the $110 \times 70m$ fifth floor of a renovated Manhattan building. G5 has about $4021m^2$ of freespace, 180 rooms, and many intersecting hallways. It is known for its ability to perplex human navigators, despite color-coded walls and art introduced as landmarks. All testing was in simulation with ROS, the state-of-the-art robot operating system (Quigley et al., 2009). MengeROS manages the simulation and deliberately introduces error into both the sensor data and action execution (Aroor et al., 2017).

To evaluate WHY we randomly sampled 5 sequences of 40 targets in each world's freespace. Table 6 reports how often voting selected each planner's submission. Two-thirds of the selected plans were based on a modified cost graph, about half of them biased by SemaFORR's spatial model. Because SemaFORR revises its model incrementally, as the robot addresses more targets, it begins to value EXPLOREP's plans less than model-based ones. For example, by the second 20 targets in each sequence of 40, plans based on the spatial model were chosen 8.2% more often, and EXPLOREP's plans 5.4% less often. No plan from NOVELP was ever selected because its plans typically performed poorly in the four affordance-based graphs. Voting, however, included NOVELP to preserve a potential trade-off between exploration and exploitation.

Table 7: Analysis of explanation results with number of unique phrasings and average readability scores

Unique phrasings	M5	H10	G5	All
Why this way?	38	30	39	49
How sure are you?	24	19	26	30
Another way?	24	19	26	30
Why yours?	17	15	16	18
How to get there?	199	175	182	556
Average readability	**M5**	**H10**	**G5**	**All**
Why this way?	4.7	5.3	5.3	5.1
How sure are you?	6.6	6.6	6.7	6.7
Another way?	3.8	2.7	3.5	3.3
Why yours?	6.8	7.0	7.2	7.0
How to get there?	7.7	7.8	7.8	7.8

We evaluated WHY for its efficiency (average computation time) and diversity (number of unique explanations produced in response to each question). We also calculated the understandability of these explanations by average reading grade level, as measured by the Coleman-Liau index (CLI) (Coleman and Liau, 1975). Since WHY's goal is to produce explanations for non-experts, lower grade-level scores are more desirable. While one could manipulate the templates to improve these scores, CLI provides a method to compare the complexity of responses to one another.

Table 7 analyzes WHY's answers to all 3000 (5 questions · 40 targets · 5 sequences · 3 worlds) questions. Its distinct natural explanations simulate people's ability to vary explanations based on context (Malle, 1999). WHY averaged 10.4 msec to compute explanations for all five questions about each plan. WHY's approach is also nuanced, with many unique responses per question. For example, WHY produced 49 unique responses to "Why does your plan go this way?" out of the 92 possible instantiations of the template. The CLI gauged them at about a sixth-grade reading level, readily understandable to a layperson.

8 Discussion

To capture useful spatial affordances for its world model, SemaFORR generalizes over its percepts, the 660 distances to the nearest obstacle that its range finder reports 15 times per second. Each of SemaFORR's planners generates paths in a graph biased by edge weights that represent that planner's objective but share an underlying structure that facilitates plan comparison. Voting guarantees that any selected plan will be optimal with respect to at least one objective, and makes it likely that the plan will also perform well with respect to the others. This also facilitates contrastive explanations in natural spatial language for the robot's planning objectives, alternative paths, and confidence.

How a robot control system represents knowledge is integral to natural communication between robots and people, especially in a spatial context. Misunderstandings between a robot and a human often arise from a discrepancy between their spatial mental models. This prompts questions about the robot's underlying decision-making and reasoning mechanisms. WHY's explanations rely on SemaFORR's cognitive underpinnings. Language about the spatial model is readily understood because SemaFORR interprets its percepts much the way people do. SemaFORR's freespace affordances were inspired by sketches after human subjects had actively explored complex virtual worlds (Chrastil and Warren, 2013). The planners' objectives are also analogous to processes empirically identified in people (Hölscher et al., 2009). The results here demonstrate that natural language communication with robots benefits substantially when a robot's control system and a human have similar cognitively-based spatial representations.

WHY's templates flexibly and quickly produce many different explanations in natural language. The templates focus language generation on SemaFORR's computational rationale rather than on linguistic structure and grammar. They also facilitate the introduction of new planners without the need to retrain a language generator for a new planning objective. For example, an objective that relied on landmarks could modify the cost graph to reduce costs near them, so that WHY might explain "I think my way is a lot better at following landmarks." Although WHY assumes the human's objective is the shortest path, it can easily substitute any objective representable in a cost graph with an admissible heuristic. SemaFORR could also incorporate a planning objective learned from external demonstration (e.g., inverse reinforcement learning) if that objective were representable as increments to the cost graph's weights.

Whenever SemaFORR selects FASTP's plan here, it assumes that it shares the human's objective. Any questions about the robot's plan necessarily challenge that assumption. Presumably, the person asks because they do not recognize their objective there. WHY responds by agreement that the person's plan is the correct way to go (e.g., "Actually, I agree that we should take the shortest route."), even though the question should not have arisen. Another way to address this would be to offer an alternative plan when FASTP is selected.

Our current work examines how well human subjects understand and feel comfortable with WHY. Although SemaFORR's parameters for intervals (e.g., in Table 3) were chosen for G5 and also worked well in other worlds, humans subject evaluation will allow us to confirm or reassess these values. Human-subject studies could also help refine WHY's explanations and incorporate psychophysics and proxemics.

Future work could extend WHY for dialogue (e.g., to clarify confusion or guide navigation (Roman et al., 2020)). This could incorporate natural language generation with deep learning and facilitate queries to the person. WHY presumes that questions arise from a difference between the human's and the robot's objectives, but they could also stem from a violation of the shared target assumption. A broader system for human-robot collaboration would seek the cause of such a mismatch, use plan explanations to resolve it, and then allow the robot to adjust its responses based on feedback from its human partner. For example, given a plan P from a person or an unspecified heuristic planner, WHY could use the individual objectives in its repertoire to tease apart and then characterize how P weighted its objectives (e.g., "So distance is more important than travel time?").

Meanwhile, SemaFORR's cognitively-based spatial model supports important path planning objectives and human-friendly explanations of its behavior, intentions, and confidence. Empirical results in three large, complex, realistic worlds show that our approach produces diverse, understandable contrastive explanations in natural language.

Acknowledgments

This work was supported in part by The National Science Foundation under CNS-1625843. The authors thank Anoop Aroor for MengeROS.

References

Anoop Aroor, Susan L Epstein, and Raj Korpan. 2017. MengeROS: A Crowd Simulation Tool for Autonomous Robot Navigation. In *Proceedings of AAAI Fall Symposium on Artificial Intelligence for Human-Robot Interaction*, pages 123–125. AAAI.

Daniel Paul Barrett, Scott Alan Bronikowski, Haonan Yu, and Jeffrey Mark Siskind. 2017. Driving Under the Influence (of Language). *IEEE Transactions on Neural Networks and Learning Systems*, pages 1–16.

Tathagata Chakraborti, Sarath Sreedharan, Sachin Grover, and Subbarao Kambhampati. 2019. Plan explanations as model reconciliation–an empirical study. In *2019 14th ACM/IEEE International Conference on Human-Robot Interaction (HRI)*, pages 258–266. IEEE.

Elizabeth R Chrastil and William H Warren. 2013. Active and passive spatial learning in human navigation: Acquisition of survey knowledge. *Journal of experimental psychology: learning, memory, and cognition*, 39(5):1520.

Elizabeth R Chrastil and William H Warren. 2014. From cognitive maps to cognitive graphs. *PloS one*, 9(11):e112544.

Meri Coleman and Ta Lin Liau. 1975. A Computer Readability Formula Designed for Machine Scoring. *Journal of Applied Psychology*, 60(2):283–284.

Susan L Epstein, Anoop Aroor, Matthew Evanusa, Elizabeth I Sklar, and Simon Parsons. 2015. Learning spatial models for navigation. In *International Conference on Spatial Information Theory*, pages 403–425. Springer.

Susan L. Epstein and Raj Korpan. 2019. Planning and explanations with a learned spatial model. In *International Conference on Spatial Information Theory*, volume 142 of *LIPICS*, pages 22:1–22:20. Schloss Dagstuhl.

Brett R Fajen and Flip Phillips. 2013. Spatial perception and action. In *Handbook of spatial cognition*. American Psychological Association.

Patrick Foo, William H Warren, Andrew Duchon, and Michael J Tarr. 2005. Do humans integrate routes into a cognitive map? map- versus landmark-based navigation of novel shortcuts. *Journal of Experimental Psychology: Learning, Memory, and Cognition*, 31(2):195–215.

Maria Fox, Derek Long, and Daniele Magazzeni. 2017. Explainable planning. In *IJCAI-17 Workshop on Explainable AI (XAI)*, page 24.

Charles R Gallistel. 1990. *The organization of learning*. The MIT Press.

James J Gibson. 1977. The theory of affordances. *Perceiving, Acting, and Knowing: Toward an Ecological Psychology*, pages 67–82.

Reginald G Golledge. 1999. Human wayfinding and cognitive maps. *Wayfinding behavior: Cognitive mapping and other spatial processes*, pages 5–45.

Antoine Grea, Laëtitia Matignon, and Samir Aknine. 2018. How explainable plans can make planning faster. In *Workshop on Explainable Artificial Intelligence*, pages 58–64.

P. E. Hart, N. J. Nilsson, and B. Raphael. 1968. A formal basis for the heuristic determination of minimum cost paths. *IEEE Transactions on Systems Science and Cybernetics*, 4(2):100–107.

Jörg Hoffmann and Daniele Magazzeni. 2019. Explainable AI planning (XAIP): overview and the case of contrastive explanation. *Reasoning Web. Explainable Artificial Intelligence*, pages 277–282.

Christoph Hölscher, Simon J Büchner, Tobias Meilinger, and Gerhard Strube. 2009. Adaptivity of wayfinding strategies in a multi-building ensemble: The effects of spatial structure, task requirements, and metric information. *Journal of Environmental Psychology*, 29(2):208–219.

Sandy H Huang, David Held, Pieter Abbeel, and Anca D Dragan. 2019. Enabling robots to communicate their objectives. *Autonomous Robots*, 43(2):309–326.

Ioannis Kostavelis and Antonios Gasteratos. 2015. Semantic mapping for mobile robotics tasks: A survey. *Robotics and Autonomous Systems*, 66:86–103.

Benjamin Krarup, Michael Cashmore, Daniele Magazzeni, and Tim Miller. 2019. Model-based contrastive explanations for explainable planning. In *ICAPS 2019 Workshop on Explainable AI Planning (XAIP)*.

Benjamin Kuipers. 2000. The spatial semantic hierarchy. *Artificial intelligence*, 119(1-2):191–233.

Christian Landsiedel, Verena Rieser, Matthew Walter, and Dirk Wollherr. 2017. A Review of Spatial Reasoning and Interaction for Real-World Robotics. *Advanced Robotics*, 31(5):222–242.

Maurício Cecílio Magnaguagno, Ramon Fraga Pereira, Martin Duarte Móre, and Felipe Rech Meneguzzi. 2017. Web planner: A tool to develop classical planning domains and visualize heuristic state-space search. In *2017 Workshop on User Interfaces and Scheduling and Planning*.

Bertram F Malle. 1999. How People Explain Behavior: A New Theoretical Framework. *Personality and Social Psychology Review*, 3(1):23–48.

Matthew Marge, Carol Espy-Wilson, and Nigel Ward. 2020. Spoken language interaction with robots: Research issues and recommendations. *Report from the NSF Future Directions Workshop*.

Raija Markkanen and Hartmut Schröder. 1997. *Hedging and Discourse: Approaches to the Analysis of a Pragmatic Phenomenon in Academic Texts*, volume 24. Walter de Gruyter.

Tim Miller. 2019. Explanation in artificial intelligence: Insights from the social sciences. *Artificial intelligence*, 267:1–38.

Van Nguyen, Stylianos Loukas Vasileiou, Tran Cao Son, and William Yeoh. 2020. Explainable planning using answer set programming. In *Proceedings of the International Conference on Principles of Knowledge Representation and Reasoning*, volume 17, pages 662–666.

Brandon S Perelman, Arthur W Evans III, and Kristin E Schaefer. 2020. Where do you think you're going? characterizing spatial mental models from planned routes. *ACM Transactions on Human-Robot Interaction (THRI)*, 9(4):1–55.

Vittorio Perera, Sai P Selveraj, Stephanie Rosenthal, and Manuela Veloso. 2016. Dynamic Generation and Refinement of Robot Verbalization. In *25th IEEE International Symposium on Robot and Human Interactive Communication (RO-MAN)*, pages 212–218. IEEE.

Thomas Pouncy, Pedro Tsividis, and Samuel J Gershman. 2021. What is the model in model-based planning? *Cognitive Science*, 45(1):e12928.

Morgan Quigley, Ken Conley, Brian P. Gerkey, Josh Faust, Tully Foote, Jeremy Leibs, Rob Wheeler, and Andrew Y. Ng. 2009. ROS: an open-source Robot Operating System. In *ICRA Workshop on Open Source Software*.

Homero Roman Roman, Yonatan Bisk, Jesse Thomason, Asli Celikyilmaz, and Jianfeng Gao. 2020. Rmm: A recursive mental model for dialog navigation. In *Proceedings of the 2020 Conference on Empirical Methods in Natural Language Processing: Findings*, pages 1732–1745.

Avi Rosenfeld and Ariella Richardson. 2019. Explainability in human–agent systems. *Autonomous Agents and Multi-Agent Systems*, 33(6):673–705.

Stephanie Rosenthal, Sai P Selvaraj, and Manuela Veloso. 2016. Verbalization: Narration of Autonomous Mobile Robot Experience. In *Proceedings of the International Joint Conference on Artificial Intelligence*, volume 16, pages 862–868.

Rosario Scalise, Stephanie Rosenthal, and Siddhartha Srinivasa. 2017. Natural Language Explanations in Human-Collaborative Systems. In *Proceedings of the 2017 ACM/IEEE International Conference on Human-Robot Interaction*, pages 377–378. ACM.

Bastian Seegebarth, Felix Müller, Bernd Schattenberg, and Susanne Biundo. 2012. Making hybrid plans more clear to human users-a formal approach for generating sound explanations. In *Proceedings of the International Conference on Automated Planning and Scheduling*, volume 22.

Avinash Kumar Singh, Neha Baranwal, Kai-Florian Richter, Thomas Hellström, and Suna Bensch. 2021. Verbal explanations by collaborating robot teams. *Paladyn, Journal of Behavioral Robotics*, 12(1):47–57.

Roykrong Sukkerd, Reid Simmons, and David Garlan. 2020. Tradeoff-focused contrastive explanation for MDP planning. In *2020 29th IEEE International Conference on Robot and Human Interactive Communication*, pages 1041–1048. IEEE.

Barbara Tversky. 1993. Cognitive maps, cognitive collages, and spatial mental models. In *European Conference on Spatial Information Theory*, pages 14–24. Springer.

William H Warren, Daniel B Rothman, Benjamin H Schnapp, and Jonathan D Ericson. 2017. Wormholes in virtual space: From cognitive maps to cognitive graphs. *Cognition*, 166:152–163.

Interactive Reinforcement Learning for Table Balancing Robot

Haein Jeon
Artificial Intelligence Robot Laboratory
Kyungpook National University
haeinjeon.knu@gmail.com

Yewon Kim
Artificial Intelligence Robot Laboratory
Kyungpook National University
yewonkim.knu@gmail.com

Boyeong Kang
Artificial Intelligence Robot Laboratory
Kyungpook National University
kby09@knu.ac.kr

Abstract

With the development of robotics, the use of robots in daily life is increasing, which has led to the need for anyone to easily train robots to improve robot use. Interactive reinforcement learning(IARL) is a method for robot training based on human–robot interaction; prior studies on IARL provide only limited types of feedback or require appropriately designed shaping rewards, which is known to be difficult and time consuming. Therefore, in this study, we propose interactive deep reinforcement learning models based on voice feedback. In the proposed system, a robot learns the task of cooperative table balancing through deep Q-network using voice feedback provided by humans in real time, with automatic speech recognition(ASR) and sentiment analysis to understand human voice feedback. As a result, an optimal policy convergence rate of up to 96% was realized, and performance was improved in all voice feedback-based models.

1 Introduction

Service robots equipped with artificial intelligence technology are increasing in daily life. Examples include museum exhibition guide robot(Thrun et al., 1999), café-serving robot(Maxwell et al., 1999), and object carrying robot(Yokoyama et al., 2003). Robots increasingly perform tasks instead of or together with humans in various environments in daily life, and there has been an active research on robots that cooperate with humans(Calinon and Billard, 2007; Du et al., 2018).

Reinforcement learning (RL) —a robot learning technique– is a method in which an agent robot learns the action of obtaining maximum rewards through trial and error. In RL, rewards are generally given by agent action in a state, and if rewards are given through real-time human-agent interaction, it is called interactive reinforcement learning(IARL). Reward shaping(RS)(Ng et al., 1999)—an IARL method—is a technique in which a human trainer modifies reward functions by providing positive or negative feedback on the action of RL agents. In previous studies on IARL using natural language, the type of feedback is very limited using fewer than 10 feedbacks(Cruz et al., 2015; Tenorio-Gonzalez et al., 2010).To facilitate the use of robots, the need for a training system through various feedbacks is raised so that robot training can be naturally performed using various voice feedbacks.

Therefore, in this study, we propose an interactive deep RL model based on voice feedback to facilitate robot use. In the proposed system, a robot uses deep Q-networks(DQNs)(Mnih et al., 2013) to perform table balancing(Kim and Kang, 2020) tasks that require cooperation with humans and learns the RL policy through RS by human voice feedback. Using RS, a human trainer who collaborates table balancing task with robot and knows how to perform a task provides positive or negative feedback in real time about a robot's action via speech. Therefore, the agent provided with voice feedback learns the optimal policy—a policy that always leads to the balanced table state—faster and more naturally than when feedback is not used.

The rest of the paper is organized as follows. Section 2 explores the flow and limitations of prior IARL studies through related work, and Section 3 describes the proposed interactive deep RL system based on voice feedback. In Section 4, we describe the results of table balancing task training based on the proposed system, and compare the difference in learning performance against conventional DQN as a baseline and between voice feedback provision types. Finally, Section 5 concludes this study and suggests future research directions.

Proceedings of Second International Combined Workshop on Spatial Language Understanding andGrounded Communication for Robotics, pages 71–78
August 5–6, 2020.

2 Related Work

One of the strategies to improve learning performance in RL is that humans guide agents as external trainers. Representative examples include learning by imitation(Bandera et al., 2012), demonstration(Argall et al., 2009; Zhu and Hu, 2018), and by feedback. Among them, focusing on feedback-providing learning, we examine: (1) the design of IARL platforms that provide feedback through mouse or remote controls (Thomaz et al., 2006; Ullerstam and Mizukawa, 2004), (2) design of IARL algorithms(Knox and Stone, 2009; Griffith et al., 2013; Faulkner et al., 2020) and (3) studies of IARL through voice feedback(Tenorio-Gonzalez et al., 2010; Cruz et al., 2015). What these studies have in common is that RS reduces training time and fosters the robot or computer to learn the target action.

Regarding methods that adopt hardware input devices, some approaches use a mouse or remote control to design an IARL platform(Thomaz et al., 2006; Ullerstam and Mizukawa, 2004). Thomaz et al. (2006) revealed that IARL can improve robot's learning efficiency in an interactive Q-learning platform for cooking simulation robots, where humans can use mouse scrolls to provide feedback for robot actions by giving a number between -1 and +1. In the study of Ullerstam and Mizukawa (2004), AIBO robots learned action sequences such as singing after hearing a command from a human feedback given by remote control. However, in these prior studies on the design of such an IARL platform, input hardware, such as a mouse and remote control, is required to provide human feedback, which is difficult to see as a natural interaction with human.

Studies on developing IARL algorithms using human feedback include TAMER (Knox and Stone, 2009), Advise (Griffith et al., 2013) and REPaIR algorithm (Faulkner et al., 2020). In TAMER—an interactive reinforcement learning algorithm proposed by Knox and Stone (2009)—an agent learns a human feedback function by receiving two evaluation signals of positive and negative from the human on their keyboards; it was tested in Tetris game and mountain car problem. In Advise proposed by Griffith et al. (2013), a human modifies an agent's action choice probability, i.e., the policy, by giving the agent binary feedback—positive or negative. As a result, Advise outperformed conventional RL algorithms on game tasks such as Pac-Man. Faulkner et al. (2020) proposed the REPaIR algorithm, which estimates the correctness of human feedback over time; virtual and physical robots performed tasks, such as putting a ball into the box in a simulation environment and grasping cup in the real world. They proved that the REPaIR algorithm matched or improved the performance of conventional Q-learning algorithms. However, these approachs that focused on feedback learning algorithms for IARL required the design of an appropriate shaping function, and additional time to calculate rewards or policies. Moreover, in the framework proposed in this study, natural language voice feedback is directly integrated into a reward so that the amount of additional computation required for DQN learning is relatively small.

Studies that investigated IARL using natural language speech voice feedback itself include dynamic RS (Tenorio-Gonzalez et al., 2010) and IARL through speech guidance(Cruz et al., 2015). Tenorio-Gonzalez et al. (2010) showed that robots can use human voice feedback in RL to learn navigation tasks by assigning specific scalar rewards to feedback vocabulary, such as +100 to "excellent" and -10 to "bad" in simulation environments. Cruz et al. (2015) used voice commands and automatic speech recognition(ASR) to transcribe input voice commands, and then compared the input sentence and predefined lists using Levenshtein distance for cleaning tasks of robot arm agents. However, in these approaches using voice feedback, the RS function was designed by assigning a static reward value to a list of very limited words and sentences defined in advance. Therefore, when a feedback vocabulary that has not been defined in the list is input, the agent may have difficulty in learning. Moreover, the framework proposed in this study analyzes the positive and negative degrees of input voice feedback using a pretrained sentiment analysis module and converts it into a reward value. Therefore, no matter what feedback phrase is input, the sentiment polarity of voice feedback can be analyzed and used for DQN RL.

Through the examination of prior studies, we can summarize that IARL ordinarily improves learning performance. However, most studies did not adopt a natural interaction method with humans by requiring hardware input devices such as a keyboard or mouse. Further, studies using voice feedback used a small number of feedbacks. In this current study, we designed an IARL system for natural

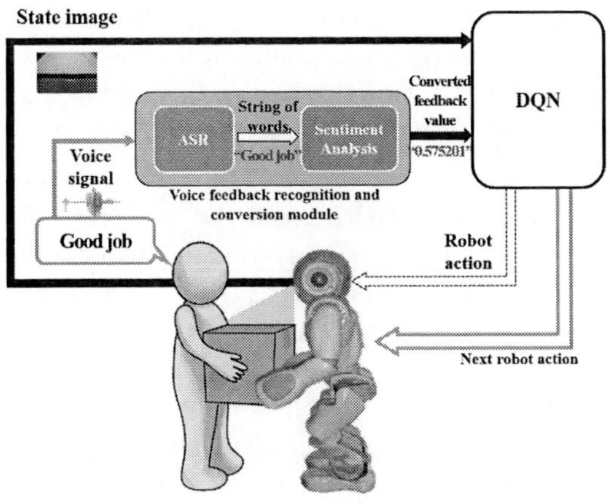

Figure 1: Interactive deep reinforcement learning model for table balancing based on human voice feedback

robot learning using voice feedback with ASR and sentiment analysis techniques to resolve these limitations.

3 Proposed Method

In this section, we describe the proposed deep RL framework for table balancing robots based on voice feedback. The task that the robot aims to learn is to maintain balance when lifting a table cooperatively with a human. Figure 1 shows the overall work diagram of the proposed system.

First, the robot takes a table state image with a camera and forwards it to the DQN. Next,the robot drives the balancing action predicted by DQN through image analysis. Then, the robot receives evaluative feedback from humans on the executed action; the voice feedback is input via the robot's microphone, converted to numerical values by voice feedback recognition and conversion module, and then incorporated into the environmental rewards of the DQN algorithm. Through repetition of the above process, the robot learns a policy in which the sum of environmental rewards and human voice feedback are maximized, and because of the learning, the robot can perform a cooperative table balancing task. In this work, the robot that will learn the table balancing task is Softbank's NAO robot, and the table is a rectangular box with width, length, and height of 31, 23, 6cm respectively. In addition, the table states to be used for learning were imaged using the lower camera mounted on the NAO robot.

Algorithm 1 Interactive Deep Q-Network Based on Voice Feedback

Initialize action-value function with random weights θ
Initialize target action-value function $\hat{Q}$ with random weights $\theta^- = \theta$
for $episodes = 1, 20000$ **do**
  Initialize sequence
  for $t = 1, T$ **do**
    Get table state image $s_t = x_t$
    With probability ϵ select a random action a_t
    Otherwise select $a_t = \text{argmax}_{a \in A} Q_t(s_t, a_t)$
    Execute action a_t and observe reward r_t and image x_{t+1}
    if Human trainer provides voice feedback f_t on state s_t **then**
      Let $r_t \leftarrow r_t + f_t$
    end if

$$y_t = \begin{cases} r_t & \text{if episode done at step } t+1 \\ r_t + \gamma \max_{a' \in A} \hat{Q}(s', a'; \theta^-)) & \text{otherwise} \end{cases} \quad (1)$$

    Perform a gradient descent step on

$$L(\theta) = \mathbb{E}[(y_t - Q(s_t, a_t; \theta_t))^2]$$

    with respect to the network parameters θ
    Every 5 steps reset $\theta^- = \theta$
  end for
end for

3.1 Deep Reinforcement Learning Process Based on Voice Feedback

The robot in the proposed system uses the DQN to recognize the table state image and output the table balancing action based on human voice feedback. A DQN combines Q-learning with a deep convolutional neural network to estimate a state–action value function (Q function) given an input image and action.

Depending on the degree of raising and balancing state of the table, the human action states are divided into five in our system: up (s_{up}), keep (s_0), down (s_{down}), up a lot(s_{upup}) and down a lot $(s_{downdown})$. The subscripts of s represent human actions. The robot executes the table balancing action a by adjusting the knee joint drive value. Five robot actions are defined depending on the direction and degree of table movement: a_{up}, a_{up}, a_0, a_{down}, and a_{down}.

Algorithm 1 represents the training process of an interactive DQN based on voice feedback. This training process is identical to the DQN training process;an interactive voice feedback-based process is added after the robot action operation. The input state s is a table image(x_t),which is an RGB image of 128×170 size representing the balance status of the table imaged by the robot camera.

Agent action	Reward
Reaching the target state	+0.5
Returning undefined action	-0.5
Reaching non-target states	-0.3

Table 1: DQN environmental reward model.

The environment selects a table state image from the training dataset and feeds it to the robot, which is a DQN agent. The robot determines the action in the current time step according to the ϵ-greedy policy, which selects a random action with a probability of ϵ for exploration. If no random action is selected, the agent chooses the action that maximizes the value of the Q function. The Q function that DQN aims to predict is as follows:

$$Q_{\pi}(s,a) = \mathbb{E}_{\pi}\sum_{t=1}^{\infty}\gamma^t r_t \quad (2)$$

where r is the reward that the robot receives when it moves to the next state from the current state by performing the action. The Q function is represented as the expected value of the cumulative reward received when executing the action a in state s, and γ is the discount rate, which reduces the influence of the Q value in the future state.

After executing an action, the agent receives evaluative voice feedback from human and environmental rewards. Table 1 defines the environmental rewards of the proposed system. The environment provides a positive reward of +0.5 when the robot reaches the target state, the balancing maintenance state (s_0). A negative reward of -0.3 is given when the agent outputs an action that reaches a state other than the target. Finally the agent receives negative reward of -0.5 when returning an undefined action other than the one in the balancing task model in Kim and Kang (2020)'s work, such as returning a_{down} while recognizing the human action state as s_{upup}.

Interactive voice feedback is a human speech evaluation of the robot's action. After checking the balance state of the table that has changed by the robot's action, the human provides positive voice feedback when the robot reaches the target state, and negative voice feedback otherwise. The provided voice feedback is converted into a numerical value through the voice feedback recognition and conversion module, and then added to the RL environment rewards. When the human provides voice feedback, the robot uses both feedback and environmental reward; and without feedback, the robot uses only environmental reward for learning. In Subsection 3.2, the voice feedback recognition and conversion module is described in depth.

In Algorithm 1, θ stands for the parameters of neural networks. DQN considers y_t as a target and proceeds learning in a direction that reduces the error of y_t and estimated $Q(s_t, a_t)$ by neural networks. Therefore, the DQN model is updated in every episode via the loss function $L(\theta)$, which computes the mean squared error. With a repetitive update of θ in the direction of minimizing $L(\theta)$, the Q function gets closer to the optimal state-action value function, and the agent learns the optimal action in the given state. Through this process, the robot can train DQN for table balancing with human voice feedback. To incorporate voice feedback in the DQN framework, we implemented voice feedback recognition and conversion module.

3.2 Voice Feedback Recognition and Conversion Modules

The voice feedback recognition and conversion module analyzed whether input voice feedback evaluated the robot's action positively or negatively. The voice feedback recognition and conversion module, shown in Figure 1, consisted of two processes: ASR and sentiment analysis.

First, the robot received an voice feedback signal from the microphone. ASR transcribed the signal into a character string and output it. We adopted Google Cloud speech-to-text as the ASR system, a cloud-based service that supported speech input and corresponding transcription in real time. This ASR system supports online streaming and offline voice audio processing, which was suitable for the agent's learning environment in our experimental setting.

Using a string of sentences obtained through ASR, sentiment analysis identified the positive and negative degrees of voice feedback phrases. The analyzed sentiment was returned in real value between -1 and 1 with positive and negative feedback being closer to +1 and -1. Moreover, if ASR could not correctly recognize speech signal, this module takes feedback as 'none' and only uses environmental reward. Google Natural Language API was used for sentiment analysis because of the ease of processing and modifying the sentiment analysis results in the implementation process.

Feedback phrases	Converted value
Well done	0.8
Fine	0.6
That is not how you do it	−0.699
Try again	−0.5

Table 2: Examples of feedback phrase with converted numeric value.

4 Experimental Results

In this section, we discuss the construction of a feedback dataset for the experiment, evaluation of the voice feedback recognition and conversion module, and verification of the proposed interactive deep RL model through experiments.

4.1 Voice Feedback Dataset and Recognition Rate

First, we constructed the voice feedback phrase dataset to test the proposed DQN model from corpora. The corpora used to build the dataset were Sentiment lexicon (Hu and Liu, 2004), AFINN lexicon (Nielsen, 2011), and Classroom English(Hong and Sohn, 2013). A total of 100 feedback dataset phrases were extracted for experiments from the corpora, with 50 positive feedback phrases and 50 negative feedback. The feedback phrases were mainly short sentences or words that evaluated actions. Table 2 shows an example of some feedback phrases in the dataset and their converted sentiment analysis values which were incorporated in the RL reward function.

As a result of testing the recognition accuracy of Google Cloud speech-to-text, which is the ASR used in this study, the average sentence recognition rate was 86% using the built feedback phrase dataset. Three times of tests with the feedback phrase datasets on Google Natural Language APIs showed an average sentence recognition rate of 96%. An accuracy of less than 100% meant that the agent might receive an erroneous reward signal due to the malfunction of the voice feedback recognition and conversion module. In this study, all cases in which wrong rewards were given from malfunction of ASR or sentiment analysis were considered, and it was confirmed via experiments that using interactive voice feedback could foster the agent's target task learning despite such errors.

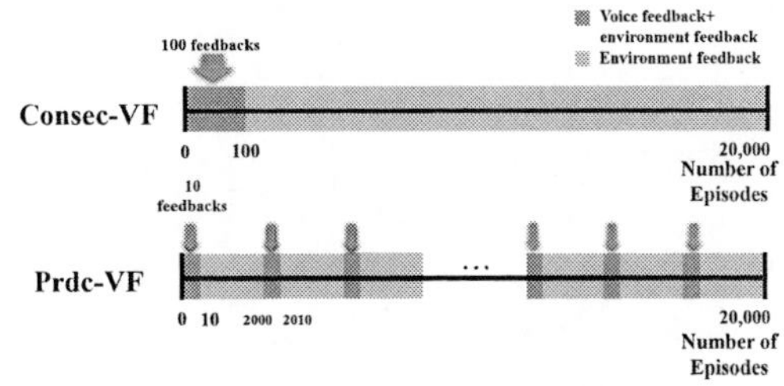

Figure 2: Comparison of Consec-VF and Prdc-VF model

Parameter	Value
Learning rate α	0.001
Discount factor γ	0.9
Epsilon ϵ	20
Number of episodes	20,000
Number of voice feedbacks	100

Table 3: Hyperparameters of DQN training.

4.2 Interactive Voice Feedback DQN Model

In this paper, we employed two voice feedback models: consecutive voice feedback (Consec-VF) and periodical voice feedback (Prdc-VF) models (Figure 2). During the training, the human can provide (1) Consec-VF in the early stages of learning, or (2) Prdc-VF throughout learning. Consec-VF provided 100 consecutive feedback earlier in training, and Prdc-VF provided 10 feedbacks every 2,000 episodes. Training was conducted in simulation where random state images are given in every episode and human trainer provides voice feedback via microphone while observing the next state. We also run experiments on a physical NAO robot as a proof of concept, and robot training video can be found at this link. (http://air.knu.ac.kr/index.php/evolutionary-cooperative-robot-development-using-distributed-deep-reinforcement-learning) We compared the two feedback-providing models with conventional DQN without voice feedback as a baseline. Additional four optimizer comparison experiments were conducted on Consec-VF.

We conducted 30 experiments for each model setting and evaluated the performance by calculating the optimal policy convergence rate after the training. Hyperparameter settings for training DQNs are shown in Table 3. All hyperparameter settings, except the number of voice feedbacks, were equally applicable to both the proposed IARL model and baseline model–DQNs.

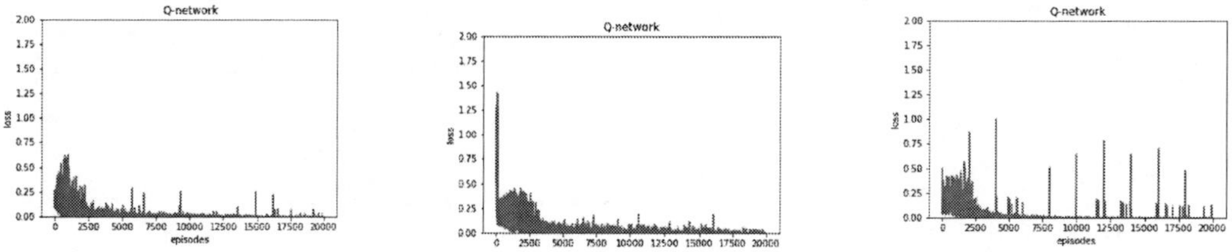

(a) Baseline (b) Consec-VF (c) Prdc-VF

Figure 3: Loss graph of models

Optimizer	Baseline	Consec-VF	Prdc-VF
SGD	80%	**86%**	80%
Adam	73 %	**96%**	60%

Table 4: Optimal policy convergence rate of 3 experimental model

Optimizer	Baseline	Consec-VF
SGD	80%	**86%**
Adam	73 %	**96%**
Adagrad	43 %	**56%**
Adadelta	63 %	**76%**

Table 5: Optimal policy convergence rate of the baseline and Consec-VF models using four different optimizers

We analyze the difference in model performance by the two methods of providing interactive voice feedback: Consec-VF and Prdc-VF. Voice feedback was provided 100 times out of 20,000 episodes (Table 3), and other episodes only used environmental rewards from Table 1. The Consec-VF model is designed to intensively feed voice feedback at the beginning of learning to establish the initial learning direction, whereas Prdc-VF model is designed to reflect human feedback steadily in the overall learning process so that human feedback could be consistently reflected.

Table 4 shows the results of experiment with two optimizers by applying the hyperparameter settings of Table 3 to the two voice feedback models and baseline DQNs. First, for the Consec-VF model, the optimal policy convergence rate was 86% and 96% when SGD and Adam optimizers were used, showing higher performance than the baseline with optimal policy convergence rates of 80% and 73% , respectively. Particularly, the convergence rate of 96% where 29 of 30 experiments learned optimal policies with Adam optimizer showed that combining Consec-VF with DQN significantly improved model performance.

Moreover, the Prdc-VF model showed lower performance than the Consec-VF and baseline models, which could be analyzed by training loss graphs. Figure 3 shows the training loss of the baseline, Consec-VF, and Prdc-VF models. In Figure 3-(a) and -(b), the loss stably converged to zero in the Consec-VF baseline model. However, in the Prdc-VF model in Figure 3-(c), loss spikes were observed during the training process. We analyzed that the intermittent intervention of voice feedback interfered with the convergence of losses during the training, resulting in a lower performance of the Prdc-VF model compared with others.

Experiment results showed that the Consec-VF model learned optimal policies better than baseline and Prdc-VF models. As in-depth experiments, we examine the results of the experiment by adding Adagrad, Adalta optimizers to the Consec-VF model to ensure that the use of Consec-VF consistently leads to model learning performance. Table 5 shows the optimal policy convergence rate after 30 experiments on the Consec-VF and baseline model on four optimizers. In all experiments Consec-VF showed improved optimal policy learning compared to the baseline DQN. These experiment results indicated that incorporating interactive voice feedback into DQN for table balancing tasks improved model learning performance in all optimizer settings.

5 Conclusion

In this study, we proposed an interactive deep RL model based on voice feedback for table balancing robot. The proposed system suggests DQN incorporating human voice feedback using ASR and sentiment analysis, where feedback given by humans are incorporated into the reward function. Experiment results show that the Consec-VF model, which provides Consec-VF early in learning, achieves an optimal policy convergence rate higher than the baseline model in all optimizer settings. There are several areas of extensions of our approach. Future direction for our work includes incorporating multimodal feedback to DQN using various robot sensors. We could also focus on deepening model optimization technique that improves learning performance of interactive RL model in varying settings. Robot could also learn when to use feedback and when to discard it or incorporate text semantics such as guiding robot behavior.

Acknowledgments

This research was supported by the National Research Foundation of Korea funded by the Korean Government under grant NRF-2019R1A2C1011270 and by the BK21 FOUR (Fostering Outstanding Universities for Research) funded by the Ministry of Education, Department of Artificial Intelligence, Kyungpook National University, Korea (I20SS7610128).

References

Brenna D Argall, Sonia Chernova, Manuela Veloso, and Brett Browning. 2009. A survey of robot learning from demonstration. *Robotics and autonomous systems*, 57(5):469–483.

JP Bandera, JA Rodriguez, L Molina-Tanco, and A Bandera. 2012. A survey of vision-based architectures for robot learning by imitation. *International Journal of Humanoid Robotics*, 9(01):1250006.

Sylvain Calinon and Aude Billard. 2007. Active teaching in robot programming by demonstration. In *RO-MAN 2007-The 16th IEEE International Symposium on Robot and Human Interactive Communication*, pages 702–707. IEEE.

Francisco Cruz, Johannes Twiefel, Sven Magg, Cornelius Weber, and Stefan Wermter. 2015. Interactive reinforcement learning through speech guidance in a domestic scenario. In *2015 international joint conference on neural networks (IJCNN)*, pages 1–8. IEEE.

G. Du, M. Chen, C. Liu, B. Zhang, and P. Zhang. 2018. Online robot teaching with natural human–robot interaction. *IEEE Transactions on Industrial Electronics*, 65(12):9571–9581.

Taylor A Kessler Faulkner, Elaine Schaertl Short, and Andrea L Thomaz. 2020. Interactive reinforcement learning with inaccurate feedback. In *2020 IEEE International Conference on Robotics and Automation (ICRA)*, pages 7498–7504. IEEE.

Shane Griffith, Kaushik Subramanian, Jonathan Scholz, Charles L Isbell, and Andrea L Thomaz. 2013. Policy shaping: Integrating human feedback with reinforcement learning. Georgia Institute of Technology.

Seonmi Hong and Jungmi Sohn. 2013. *Classroom English*. Hankukmunhwasa.

Minqing Hu and Bing Liu. 2004. Mining and summarizing customer reviews. In *Proceedings of the tenth ACM SIGKDD international conference on Knowledge discovery and data mining*, pages 168–177.

Yewon Kim and Bo-Yeong Kang. 2020. Cooperative robot for table balancing using q-learning. *The Journal of Korea Robotics Society*, 15(4):404–412.

W Bradley Knox and Peter Stone. 2009. Interactively shaping agents via human reinforcement: The tamer framework. In *Proceedings of the fifth international conference on Knowledge capture*, pages 9–16.

Bruce A Maxwell, Lisa A Meeden, Nii Addo, Laura Brown, Paul Dickson, Jane Ng, Seth Olshfski, Eli Silk, and Jordan Wales. 1999. Alfred: The robot waiter who remembers you. In *Proceedings of AAAI workshop on robotics*, pages 1–12. AAAI Press.

Volodymyr Mnih, Koray Kavukcuoglu, David Silver, Alex Graves, Ioannis Antonoglou, Daan Wierstra, and Martin Riedmiller. 2013. Playing atari with deep reinforcement learning. *arXiv preprint arXiv:1312.5602*.

Andrew Y Ng, Daishi Harada, and Stuart Russell. 1999. Policy invariance under reward transformations: Theory and application to reward shaping. In *Icml*, volume 99, pages 278–287.

Finn Årup Nielsen. 2011. A new anew: Evaluation of a word list for sentiment analysis in microblogs. *arXiv preprint arXiv:1103.2903*.

Ana C Tenorio-Gonzalez, Eduardo F Morales, and Luis Villasenor-Pineda. 2010. Dynamic reward shaping: training a robot by voice. In *Ibero-American conference on artificial intelligence*, pages 483–492. Springer.

Andrea Lockerd Thomaz, Cynthia Breazeal, et al. 2006. Reinforcement learning with human teachers: Evidence of feedback and guidance with implications for learning performance. In *Aaai*, volume 6, pages 1000–1005. Boston, MA.

S. Thrun, M. Bennewitz, W. Burgard, A. B. Cremers, F. Dellaert, D. Fox, D. Hahnel, C. Rosenberg, N. Roy, J. Schulte, and D. Schulz. 1999. Minerva: a second-generation museum tour-guide robot. In *Proceedings 1999 IEEE International Conference on Robotics and Automation (Cat. No.99CH36288C)*, volume 3, pages 1999–2005 vol.3.

Mans Ullerstam and Makoto Mizukawa. 2004. Teaching robots behavior patterns by using reinforcement learning: how to raise pet robots with a remote control. In *SICE 2004 Annual Conference*, volume 1, pages 143–146. IEEE.

Kazuhiko Yokoyama, Hiroyuki Handa, Takakatsu Isozumi, Yutaro Fukase, Kenji Kaneko, Fumio Kanehiro, Yoshihiro Kawai, Fumiaki Tomita, and Hirohisa Hirukawa. 2003. Cooperative works by a human and a humanoid robot. In *2003 IEEE International Conference on Robotics and Automation (Cat. No. 03CH37422)*, volume 3, pages 2985–2991. IEEE.

Zuyuan Zhu and Huosheng Hu. 2018. Robot learning from demonstration in robotic assembly: A survey. *Robotics*, 7(2):17.

Multi-Level Gazetteer-Free Geocoding

Sayali Kulkarni[*]
Google Research
sayali@google.com

Shailee Jain[*†]
University of Texas, Austin
shailee@cs.utexas.edu

Mohammad Javad Hosseini[†]
University of Edinburgh
javad.hosseini@ed.ac.uk

Jason Baldridge
Google Research
jasonbaldridge@google.com

Eugene Ie
Google Research
eugeneie@google.com

Li Zhang
Google Research
liqzhang@google.com

Abstract

We present a multi-level geocoding model (MLG) that learns to associate texts to geographic coordinates. The Earth's surface is represented using space-filling curves that decompose the sphere into a hierarchical grid. MLG balances classification granularity and accuracy by combining losses across multiple levels and jointly predicting cells at different levels simultaneously. It obtains large gains without any gazetteer metadata, demonstrating that it can effectively learn the connection between text spans and coordinates—and thus makes it a gazetteer-free geocoder. Furthermore, MLG obtains state-of-the-art results for toponym resolution on three English datasets without any dataset-specific tuning.

1 Introduction

Geocoding is the task of resolving location references in text to geographic coordinates or regions. It is often studied in social networks, where metadata and the network itself provide additional non-textual signals (Backstrom et al., 2010; Rahimi et al., 2015). If locations can be mapped to an entity in a knowledge graph, toponym resolution – a special case of entity resolution – can be used to resolve references to locations. Past work used heuristics based on location popularity (Leidner, 2007) and distance between candidate locations (Speriosu and Baldridge, 2013), as well as learned associations from text to locations. However, such approaches have a strong bias for highly-populated locations, especially for social media.

We present Multi-Level Geocoder (MLG, Fig. 1), a model that learns spatial language representations and maps toponyms to coordinates on Earth's surface. This geocoder is not restricted to resolving toponyms to specific location *entities*, but rather to geo-coordinates directly. MLG can thus be extended to any arbitrary location references in future without having to rely on its presence in the gazetteer. For comparative evaluation, we use three English toponym resolution datasets from

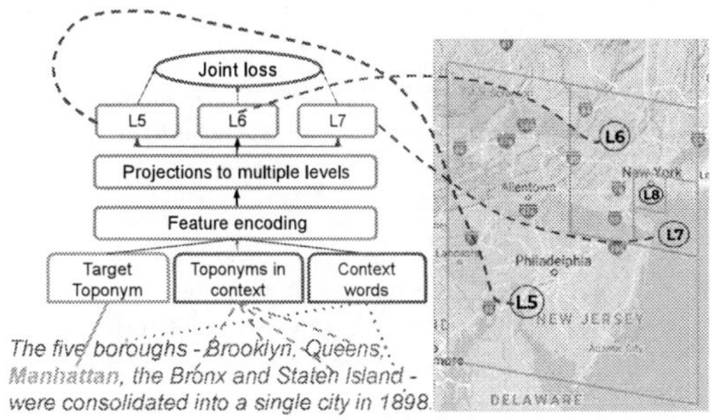

Figure 1: Overview of Multi-Level Geocoder, using multiple context features and jointly predicting cells at multiple levels of the S2 hierarchy.

distinct textual domains. MLG shows strong performance, even without gazetteer and population metadata.

MLG is a text-to-location neural geocoder. We represent the locations using S2 geometry[1]—a hierarchical discretization of the Earth's surface based on space-filling curves. S2 naturally supports spatial representation at multiple levels, including very fine grained cells (as small as 1cm^2 at level 30). Here, we use combinations of levels 4 ($\sim$300K km^2) to 8 ($\sim$1K km^2). Large cells are easy to predict accurately; however, they are too coarse on their own, and perform poorly on metrics that consider error distances. Smaller cells improve granularity but result in larger and harder output spaces with less training evidence per cell. MLG balances classification granularity and accuracy by predicting at multiple S2 levels and jointly optimizing for the loss at each level. Fig. 1 shows an area around New York City covered by cell id `0x89c25` at level 8 and `0x89c4` at level 5. This is more fine-grained than previous work that does text-to-location geocoding (Gritta et al., 2018a), which uses arbitrary square-degree cells, e.g. $2°$-by-$2°$ cells ($\sim$48K km^2).

Unlike previous work that relies on external gazetteer information, MLG is more flexible and can predict geolocation only from context. For instance, it predicts the location of *Manhattan* from the surrounding words (*The five boroughs - Brooklyn, Queens, the Bronx and*

[*] Equal contribution
[†] Work done during internship at Google

[1] `https://s2geometry.io/`

Proceedings of Second International Combined Workshop on Spatial Language Understanding andGrounded Communication for Robotics, pages 79–88
August 5–6, 2020. ©2021 Association for Computational Linguistics

Staten Island - . . .). Earlier approaches instead relied on a knowledge graph that had *Manhattan* as an entity. While the hierarchical geolocation model of Wing and Baldridge (2014) over *kd*-trees has some more fine-grained cells, MLG predicts over a much larger set of smaller cells. Furthermore, MLG is a single model that jointly incorporates multiple levels rather than ensembling independent per-cell models for each level.

Our main contributions are the following.

- We define MLG, a model that jointly predicts cells at multiple levels, including finer-grained cells than previous work.
- We show that S2 provides a strong and standardized hierarchical discretization of the Earth's surface for cell-based geocoders.
- We show that it is possible and even preferable to eschew gazetteer metadata. In particular, our experiments show that this strategy generalizes much better.
- We show state-of-the-art performance on three English datasets *without* any fine-tuning.
- When analyzing these datasets, we found inconsistencies in the true coordinates that we unify to support consistent evaluation.[2]

2 Spatial representations

Geocoders map text spans to geo-coordinates—a prediction over a continuous space representing the surface of a sphere. We relax the problem from continuous space to discrete space by quantizing the Earth's surface as a grid and performing multi-class prediction over the grid's cells. We construct a hierarchical grid using the S2 library.[3] S2 projects the six faces of a cube onto the Earth's surface and each face is recursively divided into 4 quadrants, as shown in Figure 1. Cells at each level are indexed using a Hilbert curve. Each S2 cell is represented as a 64-bit unsigned integer and can correspond to areas as small as $\approx 1cm^2$. S2 cells preserve cell size across the globe better than commonly-used degree-square grids (e.g. 1°x1°) (Serdyukov et al., 2009; Wing and Baldridge, 2011). Hierarchical triangular meshes (Szalay et al., 2007) and Hierarchical Equal Area iso-Latitude Pixelation (Melo and Martins, 2015) are alternatives that preserve cell size better, but S2 is easier to work with and has strong, standard tooling.

Our experiments go as far as S2 level eight (of thirty), but our approach is extendable to any level of granularity and could support very fine-grained locations like buildings and landmarks. The built-in hierarchical nature of S2 cells makes it well suited as a scaffold for models that learn and combine evidence from multiple levels. This combines the best of both worlds: specificity at finer levels and aggregation/smoothing at coarser levels.

Roller et al. (2012) use adaptive, variable shaped cells based on k-d trees; such grids can adapt to the different shapes of a region but depend on the locations of labeled examples in a training resource. As such, a k-d tree grid may not generalize well to examples with different distributions from training resources. Spatial hierarchies based on containment relations among entities rely heavily on metadata like GeoNames (Kamalloo and Rafiei, 2018). Polygons for geopolitical entities such as city, state, and country (Martins et al., 2015) are perhaps ideal, but these too require detailed metadata for all toponyms, managing non-uniformity of the polygons, and general facility with GIS tools. The Point-to-City (P2C) method applies an iterative k-d tree-based method for clustering coordinates and associating them with cities (Fornaciari and Hovy, 2019b). S2 can represent such hierarchies in various levels without relying on external metadata.

S2 Level	number of cells	Avg area
L4	1.5k	332
L5	6.0k	83
L6	24.0k	21
L7	98.0k	5
L8	393.0k	1

Table 1: S2 levels used in MLG. Average area is in 1k km^2.

In accordance with the nature of the problem over continuous space, studies using bivariate Gaussians on multiple flattened regions (Eisenstein et al., 2010; Priedhorsky et al., 2014)) perform well on distance based metrics, but this involves difficult trade-offs between flattened region sizes and the level of distortion they introduce. Some of the early models used with grid-based representations were probabilistic language models that produce document likelihoods in different geospatial cells (Serdyukov et al., 2009; Wing and Baldridge, 2011; Dias et al., 2012; Roller et al., 2012). Extensions include domain adapting language models from various sources (Laere et al., 2014), hierarchical discriminative models (Wing and Baldridge, 2014; Melo and Martins, 2015), and smoothing sparse grids with Gaussian priors (Hulden et al., 2015). Alternatively, Fornaciari and Hovy (2019a) use a multi-task learning setup that assigns probabilities across grids and also predicts the true location through regression. Melo and Martins (2017) cover a broad survey of document geocoding. Much of this work has been conducted on social media data like Twitter, where additional information beyond the text—such as the network connections and user and document metadata—have been used (Backstrom et al., 2010; Cheng et al., 2010; Han et al., 2014; Rahimi et al., 2015, 2016, 2017). MLG is not trained on social media data and hence, does not need additional network information. Further, the data does not have a character limit like tweets, so models can learn from long text sequences.

[2] https://github.com/google-research-datasets/mlg_evaldata

[3] https://s2geometry.io/

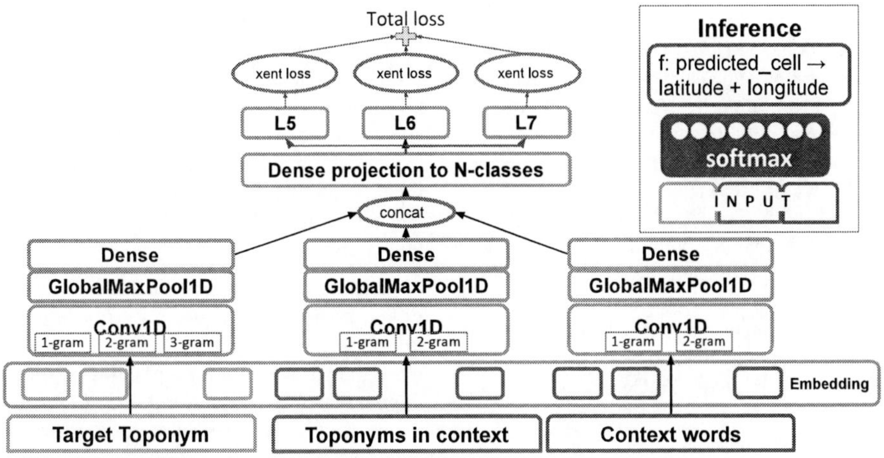

Figure 2: Multi-Level Geocoder model architecture and inference setup.

3 Multi-Level Geocoder (MLG)

Multi-Level Geocoder (MLG, Figure 2) is a text-to-location CNN-based geocoder. Context features are similar to CamCoder (Gritta et al., 2018a) but we exclude its metadata-based MapVec feature. Locations are represented using a hierarchical S2 grid; this enables joint multi-level prediction, by optimizing for total loss computed from all levels.

3.1 Prior geocoding models

Toponym resolution identifies place mentions in text and predicting the precise geo-entity in a knowledge base (Leidner, 2007; Gritta et al., 2018b). The knowledge base is then used to obtain the geo-coordinates of the predicted entity for the geocoding task. Rule-based toponym resolvers (Smith and Crane, 2001; Grover et al., 2010; Tobin et al., 2010; Karimzadeh et al., 2013) rely on hand-built heuristics like population from metadata resources like Wikipedia and GeoNames[4] gazetteer. This works well for many common places, but it is brittle and cannot handle unknown or uncommon place names. As such, machine learned approaches that use toponym context features have demonstrated better performance (Speriosu and Baldridge, 2013; Zhang and Gelernter, 2014; DeLozier et al., 2015; Santos et al., 2015). A straightforward–but data hungry–approach learns a collection of multi-class classifiers, one per toponym with a gazetteer's locations for the toponym as the classes (e.g., the WISTR model of Speriosu and Baldridge (2013)).

A hybrid approach that combines learning and heuristics by predicting a distribution over the grid cells and then filtering the scores through a gazetteer works for systems like TRIPDL (Speriosu and Baldridge, 2013) and TopoCluster (DeLozier et al., 2015). A combination of classification and regression loss to predict over recursively partitioned regions shows promising results with in-domain training (Cardoso et al., 2019). CamCoder (Gritta et al., 2018a) uses this strategy with a much stronger neural model and achieves state-of-the-art results. It incorporates side metadata in the form of its *MapVec* feature vector, which encodes knowledge of potential locations and their populations matching all toponym in the text. It thus uses population signals in both the MapVec feature in training and in output predictions biasing the predictions toward locations with larger populations.

[4] www.geonames.org

3.2 Building blocks

MLG uses a convolutional neural network to map input text to S2 cells at a given granularity.

Input MLG extracts three features from the input context: (a) token sequence ($w_{a,1:l_a}$) is all the tokens in input, (b) toponym mentions ($w_{b,1:l_b}$) is the list of all locations words in the context, and (c) surface form of the target toponym ($w_{c,1:l_c}$) that is to be geo-located. All text inputs are transformed uniformly, using shared model parameters. Let input text content be denoted as a word sequence $w_{x,1:l} = [w_{x,1}, \dots, w_{x,l}]$, initialized using GloVe embeddings $\phi(w_{x,1:l}) = [\phi(w_{x,1}), \dots, \phi(w_{x,l})]$ (Pennington et al., 2014).

Consider a short context for *Manhattan* as "*Manhattan is the smallest and most densely populated borough compared to others - Bronx, Brooklyn, Queens, and Staten Island.*" All tokens are lower cased and we get w_a as ["*is*", "*the*", "*smallest*", "*and*", ...], toponym mentions w_b are ["*bronx*", ... , "*staten*", "*island*"], and surface form of target toponym w_c would be "*manhattan*".

Model 1D convolutional filters capture n-gram sequences through $\texttt{Conv1D}_n(\cdot)$, followed by max pooling and then projection to a dense layer to get $\texttt{Dense}(\texttt{MaxPool}(\texttt{Conv1D}_n(\phi(w_{x,1:l}))))\in\mathbb{R}^{2048}$, where $n=\{1,2\}$ for the token sequence and toponym mentions, and $n=\{1,2,3\}$ for the target toponym.

These projections are concatenated to form the full input representation. MLG is designed to study effectiveness of spatial language representation without any gazetteer information. Hence we choose a CNN-based architecture, but can be extended to large scale pretrained language models (Devlin et al. (2018)).

Output An S2 cell is predicted at the highest granularity using a softmax over the output space. The center of the predicted S2 cell is taken as the predicted coordinates. *Optionally*, the predicted cells may be snapped to the closest valid cells that overlap the potential gazetteer locations for the toponym, weighted by their population (similar to previous work, like CamCoder).

3.3 Multi-level classification

MLG's core block is a multi-class classifier using a CNN. Rather than predicting cells at a single level, we project the output onto multiple levels with a multi-headed model. The penultimate layer maps representations of the input to probabilities over the finest-grained cells. Gradient updates are computed using cross entropy loss between predicted probabilities $\mathbf{p}$ and the one-hot true class vector $\mathbf{c}$.

MLG exploits the natural hierarchy of geographic locations by jointly predicting at different levels of granularity. CamCoder uses 7.8K output classes representing 2x2 degree tiles (after filtering cells that have no support in training, such as over bodies of water, to limit the class space). This requires maintaining a cumbersome mapping between actual grid cells and the classes. MLG's multi-level hierarchical representation overcomes this problem by including coarser levels (like L5) to guide the predictions at finer-grained levels. We focus on three levels that are appropriate for the task: L5, L6 and L7 (shown in Table 1), each giving 6K, 24K, and 98K output classes, respectively.

We define losses at each level (L5, L6, L7) and minimize them jointly, i.e., $\mathcal{L}_{\text{total}} = (\mathcal{L}(\mathbf{p}_{\text{L5}}, \mathbf{c}_{\text{L5}}) + \mathcal{L}(\mathbf{p}_{\text{L6}}, \mathbf{c}_{\text{L6}}) + \mathcal{L}(\mathbf{p}_{\text{L7}}, \mathbf{c}_{\text{L7}}))/3$. At inference time, a single forward pass computes probabilities at all three levels. The final score for each L7 cell is dependent on its predicted probability as well as the probabilities in its corresponding parent L6 and L5 cells. Then the final score for $s_{\text{L7}}(f) = \mathbf{p}_{\text{L7}}(f) * \mathbf{p}_{\text{L6}}(e) * \mathbf{p}_{\text{L5}}(d)$ and the final prediction is $\hat{y} = \text{argmax}_y\, s_{\text{L7}}(y)$. This approach is easily extensible to capture additional levels of resolution—we also present results with finer resolution at L8, with $\sim$1K km^2 area and coarser resolution at L4 with $\sim$300K km^2 area for comparison.

3.4 Gazetteer-constrained prediction

The only way MLG uses geographic information is from training labels for toponym targets. At test time, MLG predicts a distribution over all cells at each S2 level given the input features and picks the highest probability cell at the most granular level. We use the center of the cell as predicted coordinates. However, when the goal is to resolve a specific toponym, an effective heuristic is to use a gazetteer to filter the output predictions to only those that are valid for the toponym. Furthermore, gazetteers come with population information that can be used to nudge predictions toward locations with high populations—which tend to be discussed more than less populous alternatives. Like DeLozier et al. (2015), we consider both gazetteer-free and gazetteer-constrained predictions.

Gazetteer-constrained prediction makes toponym resolution a sub-problem of entity resolution. As with broader entity resolution, a strong baseline is an alias table (the gazetteer) with a popularity prior. For geographic data, the population of each location is an effective quantity for characterizing popularity: choosing Paris, France rather than Paris, Texas for the toponym *Paris* is a better bet. This is especially true for zero-shot evaluation where one has no in-domain training data.

We follow the strategy of Gritta et al. (2018a) for gazetteer constrained predictions. We construct an alias table which maps each mention m to a set of candidate locations, denoted by $C(m)$ using link information from Wikipedia and the population $\text{pop}(\ell)$ for each location ℓ is read from WikiData.[5] For each of the gazetteer's candidate locations we compute a population discounted distance from the geocoder's predicted location p and choose the one with smaller value as $\text{argmin}_{\ell \in C(m)} \text{dist}(p, \ell) \cdot (1 - c \cdot \text{pop}(\ell)/\text{pop}(m))$. Here, $\text{pop}(m)$ is the maximum population among all candidates for mention m, $\text{dist}(p, \ell)$ is the great circle distance between prediction p and location ℓ, and c is a constant in $[0, 1]$ that indicates the degree of population bias applied. For c=0, the location nearest the prediction is chosen (ignoring population); for c=1, the most populous location is chosen, (ignoring p). This is set to 0.9, which worked best on the development set.

3.5 Training Data and Representation

MLG is trained on geographically annotated Wikipedia pages, *excluding* all pages in WikToR (see Sec. 4.1). For each page with geo-coordinates, we consider context windows of up to 400 tokens (respecting sentence boundaries) as training example candidates. Only context windows that contain the target Wikipedia toponym are used. We use Google Cloud Natural Language API libraries to tokenize[6] the page text and for identifying[7] toponyms in the contexts. We use the July 2019 English Wikipedia dump, which has 1.11M location annotated pages giving 1.76M training examples. This is split 90/10 for training/development.

[5] http://www.wikidata.org

[6] https://cloud.google.com/natural-language/docs/analyzing-syntax

[7] https://cloud.google.com/natural-language/docs/analyzing-entities

Gaz Used	Model	AUC of error curve ↓ WTR	LGL	GV	Avg	accuracy@161 ↑ WTR	LGL	GV	Avg	Mean error ↓ WTR	LGL	GV	Avg
Yes	POPBASELINE	66	42	41	50	22	57	68	49	4175	1933	898	2335
	CAMCODER	24	32	15	24	72	63	82	72	440	877	315	544
	SLG 7	17	28	**13**	19	82	72	**86**	80	480	648	305	478
	MLG 5-7	**15**	**27**	**13**	**18**	**85**	**73**	85	**81**	**347**	**620**	**276**	**414**
No	CAMCODER	49	60	65	58	70	38	26	45	239	1419	2246	1301
	SLG 7	39	55	56	50	86	49	48	61	424	1688	1956	1356
	MLG 5-7	**37**	**54**	**55**	**49**	**91**	**53**	**49**	**64**	**180**	**1407**	**1690**	**1092**

Table 2: Comparing population baseline, CamCoder benchmark (our implementation), and our SLG and MLG models on the *unified* data, both with and without the gazetteer filter.

Inference	AUC of error curve ↓ WTR	LGL	GV	Avg	accuracy@161 ↑ WTR	LGL	GV	Avg	Mean error ↓ WTR	LGL	GV	Avg
L5-7	**37**	**54**	**55**	**49**	**91**	**53**	**49**	**64**	**180**	**1407**	**1690**	**1092**
Only L5	48	60	62	57	79	45	39	54	285	1599	1957	1280
Only L6	43	57	60	53	90	51	44	62	265	1534	2003	1267
Only L7	38	**54**	56	50	89	51	48	63	349	1525	2014	1296

Table 3: Prediction granularity: performance of MLG trained with multi-level loss on L5, L6 and L7 but using single level at inference time.

4 Evaluation

We train MLG as a general purpose geocoder and evaluate it on toponym resolution. A strong baseline is to choose the most populous candidate location (POPBASELINE): i.e. $\text{argmax}_{\ell \in C(m)} \text{pop}(\ell)$

4.1 Evaluation Datasets

We use three public datasets: Wikipedia Toponym Retrieval (WikToR) (Gritta et al., 2018b), Local-Global Lexicon (LGL) (Lieberman et al., 2010), and GeoVirus (Gritta et al., 2018a). See Gritta et al. (2018b) for extensive discussion of other datasets.

WikToR (WTR) is the largest programmatically created corpus that allows for comprehensive evaluation of toponym resolvers. By construction, ambiguous location mentions were prioritized (e.g. *"Lima, Peru"* vs. *"Lima, Ohio"* vs. *"Lima, Oklahoma"* vs *"Lima, New York"*). As such, population-based heuristics are counter-productive in WikToR.

LGL consists of 588 news articles from 78 different news sources. This dataset contains 5,088 toponyms and 41% of these refer to locations with small populations. About 16% of the toponyms are for street names, which do not have coordinates; and hence dropped from our evaluation set. About 2% have an entity that does not exist in Wikipedia, which were also dropped thus leaving 4,172 examples for evaluation.

GeoVirus (GV) is based on 229 WikiNews[8] articles about global epidemics obtained using keywords such as "Bird Flu" and "Ebola". Place mentions are manually tagged and assigned Wikipedia page URLs. In total, this dataset provides 2,167 toponyms for evaluation.

WikToR serves as in-domain Wikipedia-based evaluation data, while both LGL and GeoVirus provide out-of-domain news corpora evaluation.

[8] https://en.wikinews.org

4.2 Unified evaluation sets

We use the publicly available versions of the three datasets used in CamCoder.[9] However, after analyzing examples across all of them, we identified inconsistencies in location target coordinates.

First, WikToR's evaluation set delivers annotations based on GeoNames DB and Wikipedia APIs. We discovered that WikToR was annotated with an older version of GeoNames DB, which has a known issue of sign flip in either latitude or longitude of some locations. For example, *Santa Cruz, New Mexico* was incorrectly tagged as (35, 106) instead of (35, -106). This affects 296 out of 5,000 locations in WikToR—mostly cities in the United States and a few in Australia.

Second, the target coordinates are inconsistent across the 3 datasets. For example, Canada is (60.0, -95.0) in GeoVirus, (60.0, -96.0) in LGL and (45.4, -75.7) in WikToR. Given our point-based representations, we need consistent coordinates across the evaluation sets. So we re-annotated all three datasets to unify the coordinates for target toponyms.[2] This was done Wikidata to be consistent with Wikipedia training labels.

4.3 Evaluation Metrics

We use three metrics for evaluation: AUC for the error curve, accuracy@161km and mean distance error. AUC[10] is the area under the discrete curve of sorted log-error distances. This is captures the entire distribution of errors and is not sensitive to outliers. It uses the log of the error distances, which appropriately focuses the metric on smaller error distances. Accuracy is the percentage of toponyms that are resolved to within 161km

[9] https://github.com/milangritta/Geocoding-with-Map-Vector/tree/master/data

[10] Unlike the standard AUC, lower is better for AUC since this is based on the curve of error distances.

Model	Dev loss	AUC of error curve ↓				accuracy@161 ↑				Mean error ↓			
		WTR	LGL	GV	Avg	WTR	LGL	GV	Avg	WTR	LGL	GV	Avg
MLG 4-7	8.71	37	55	54	49	91	51	51	64	197	1529	1570	1099
MLG 5-7	**7.25**	37	54	55	49	91	53	49	64	180	1407	1690	1092
MLG 5-8	13.28	38	58	67	54	89	45	24	53	272	1866	3058	1732

Table 4: Models trained with different granularities help trade-off between accuracy and generalization. Selected model MLG 5-7 is based on optimal performance of the holdout.

(100 miles) of their true location. Mean distance error is the average of all distances between predicted locations (center of the predicted S2 cell) and true locations of the target toponym.

We study the benefits of resolving toponyms over multiple levels to account for the range of populations, resolution ambiguity, topological shapes and sizes of different toponyms. We leave the shaping of the output space as future work (e.g., using geopolitical polygons instead of points).

5 Experiments

5.1 Training

MLG is trained using TensorFlow (Abadi et al., 2016) distributed across 13 P100 GPUs. Each training batch processes 512 examples. The model trains up to 1M steps, although they converge around 500K steps. We found an optimal initial learning rate of 10^{-4} decaying exponentially over batches after initial warm-up. For optimization, we use Adam (Kingma and Ba, 2015) for stability.

We considered S2 levels 4 through 8, including single level (SLG) and multi-level (MLG) variations. MLG's architecture offers the flexibility of doing multi-level training but performing prediction with just one level. Based on the loss on Wikipedia development split, we chose multi-level training and prediction with levels 5, 6 and 7.

We stress that our focus is *geocoding without gazetteer information at inference time*. However, we also show that additional gains can be achieved using gazetteers to select relevant cells for a given toponym, and scale the output using the population bias (c) as described in section 3.4.

5.2 Results

Table 2 shows results for the POPBASELINE, CAMCODER, SLG and MLG models on all three datasets for all metrics. For CAMCODER, SLG and MLG, we include results with and without gazetteer filtering (sect. 3.4). Results are reported on the unified datasets. The CAMCODER results are from our own implementation and trained on the same examples as MLG training set.

Overall trends The most striking result is MLG's improvement over CAMCODER without gazetteer filtering, especially on WikToR—a dataset specifically designed to counteract population priors. MLG clearly generalizes better by leaving out the non-lexical MapVec feature and thereby avoiding the influence of its population bias for the toponyms in the context.

Fine-grained multi-level learning and prediction pays off, both with and without gazetteer filtering. This is particularly clear with AUC, where MLG is 6% better (averaged over all datasets) than CAMCODER with the gazetteer filter. Without the filter, MLG has an even larger gain of 9%.

Generalization When not using the gazetteer filter, MLG actually beats the population baseline for WikToR, and it is much closer to the strong population baselines for LGL and GeoVirus than CAMCODER and SLG. This indicates that the multi-level approach allows the use of training evidence to generalize better over examples drawn globally (entire world in GeoVirus) as well as locally (the United States of America in LGL).

Multi-level prediction helps. Table 3 compares performance of using individual levels from the same MLG model trained on levels L5, L6 and L7 (without the gazetteer filter). The trade off of predicting at different granularity is clear: when we use lower granularity, e.g. L5 cells, our model can generalize better, but it may be less precise given the large size of the cells. On the other hand, when using finer granularity, e.g. L7 cells, the model can be more accurate in dense regions, but could suffer in sparse regions where there is less training data. Combining the predictions from all levels balances the strengths effectively.

Levels five through seven offer best tradeoff Table 4 shows performance of MLG by training and predicting with multiple levels at different granularities. Overall, using levels five through seven (which has the best development split loss) provides the strongest balance between generalization and specificity. For locating cities, states and countries, especially when choosing from candidate locations in a gazetteer, L8 cells do not provide much greater precision than L7 and suffer from fewer examples as evidence in each cell.

Qualitative examples An effective use of context in correctly predicting coordinates is shown in Table 5 on two examples, *Arlington* and *Lincoln*. In both pairs, the context helps to shift the predictions in the right regions on the map. It is not biased by just the most populous place. Here we only show a part of the context for clarity though the actual context is longer (see Sec. 3.5).

Arlington is a former manor, village and civil parish in the North Devon district of Devon in England. The parish includes the villages of Arlington and Arlington Beccott. ...

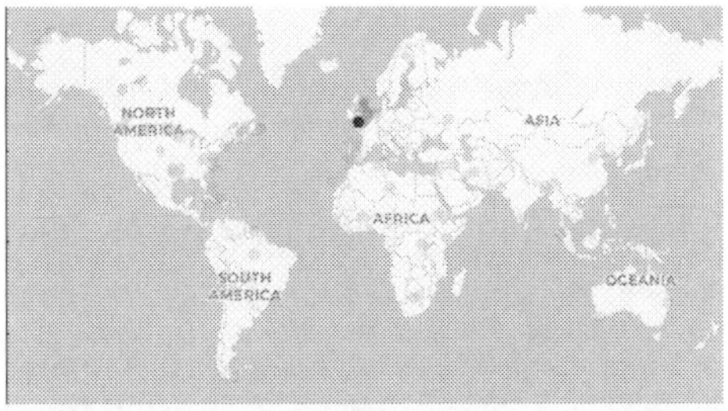

Arlington is a city in Gilliam County, Oregon, United States. The account of how the city received its name varies; one tradition claims it was named after the lawyer Nathan Arlington Cornish, ...

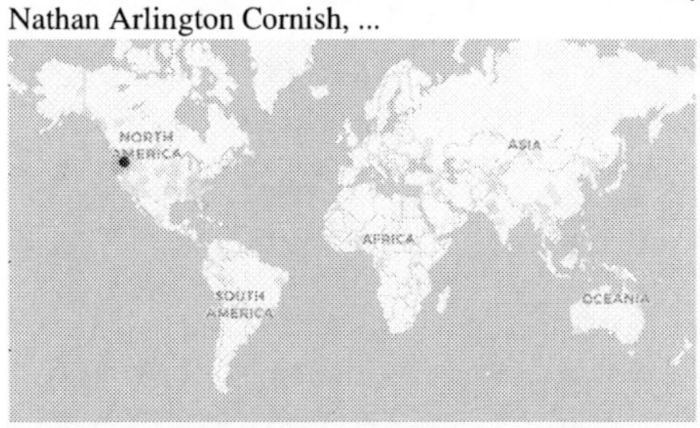

Lincoln is a city in Logan County, Illinois, United States. It is the only town in the United States that was named for Abraham Lincoln before he became president....

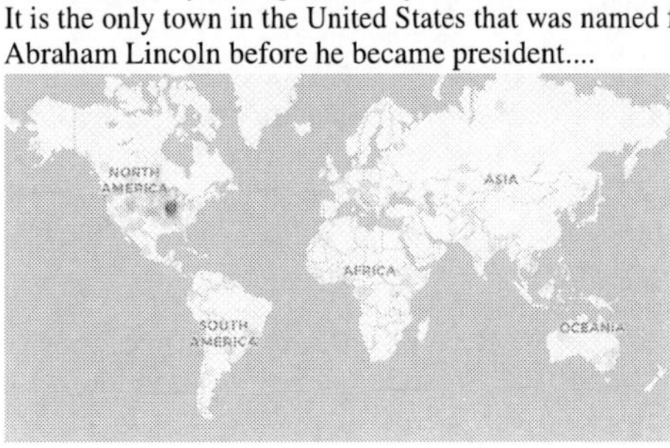

Lincoln is a city in the province of Buenos Aires in Argentina. It is the capital of the district of Lincoln (Lincoln Partido). The district of Lincoln was established on ...

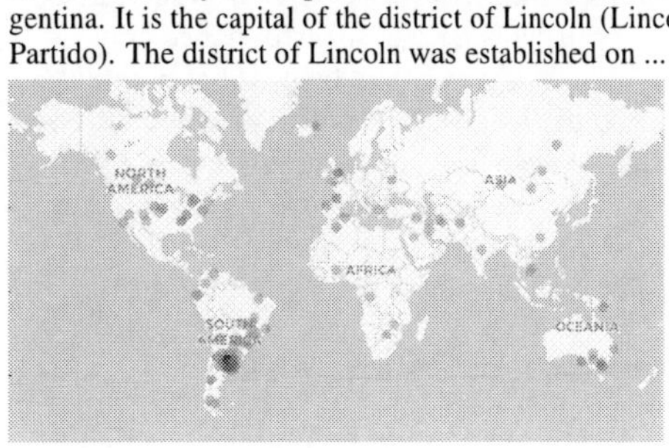

Table 5: Context – terms and other toponyms – drive the probabilities in the right regions to correctly geo-locate *Arlington* (top) and *Lincoln* (bottom) distributions in different parts of the world.

	AUC of error curve ↓				accuracy@161 ↑				Mean error ↓			
Ablation	WTR	LGL	GeoV	Avg	WTR	LGL	GeoV	Avg	WTR	LGL	GeoV	Avg
all features	37	54	55	49	91	53	49	64	180	1407	1690	1092
- target	38	60	69	55	91	39	18	49	174	2032	2811	1672
- all toponyms	69	75	82	76	29	14	4	16	4487	4442	6360	5096

Table 6: Effect of ablating location features from the input to demonstrate their importance in MLG 5-7.

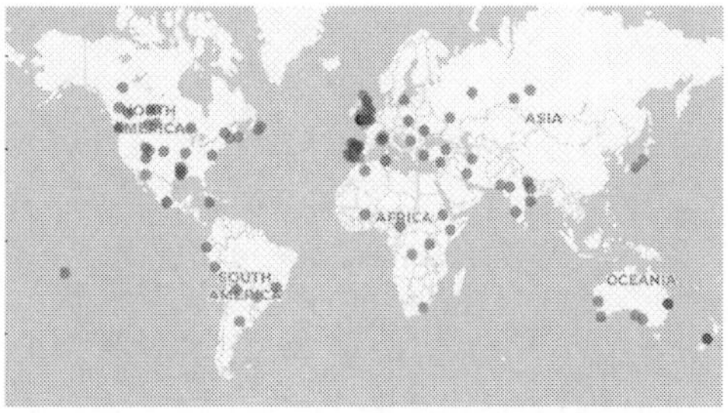

Figure 3: Ablating all toponyms at inference time spreads out the probabilities (points lighted up all over the map) but can still correctly predict *Arlington (England)* purely from context.

Ablations Table 6 shows ablation of salient features at inference time, removing either the target toponym or all toponyms. While masking the target toponym does not change results much except for GeoVirus, masking all other toponyms degrades performance considerably. Nevertheless, it may still be possible with just the context words, which include other named entities, characteristics of the place, and location-focused words in few cases. For example, *Arlington (England)* can be geolocated after all toponyms are masked (Fig. 3), though the distribution is more spread out in this case.

6 Conclusion and Future work

MLG uses multi-level optimization for the inherently hierarchical problem of geocoding. With just textual inputs, we can predict the location of a target toponym with minimal to no metadata from gazetteer and outperform existing benchmark models. MLG can thus be used as a gazetteer-free geocoder, on inputs like historical texts (DeLozier et al., 2016). Further, the models generalize very well across domains, and thus can be used in real-time datasets like news feeds. The multi-level loss can be further refined by using approaches like hierarchical softmax (Morin and Bengio, 2005) to incorporate the conditional probabilities across layers more effectively.

A natural extension would be to fine-tune large pre-trained language models for the geocoding task. We expect that the potential value of this is orthogonal to the contribution of our multi-level loss and the use of S2 cells. Another future direction involves smoothing the label space during training to capture the relations among spatial close cells by defining the loss as a function of Earth mover's distance with approximations like Sinkhorn divergence. This would also enable shaping the output class space to polygons instead of points, which is more realistic for geographical regions.

References

Martín Abadi, Paul Barham, Jianmin Chen, Zhifeng Chen, Andy Davis, Jeffrey Dean, Matthieu Devin, Sanjay Ghemawat, Geoffrey Irving, Michael Isard, et al. 2016. TensorFlow: A system for large-scale machine learning. In *12th {USENIX} Symposium on Operating Systems Design and Implementation ({OSDI} 16)*, pages 265–283.

Lars Backstrom, Eric Sun, and Cameron Marlow. 2010. Find me if you can: Improving geographical prediction with social and spatial proximity. In *Proceedings of the 19th International Conference on World Wide Web, WWW '10*, pages 61–70.

Ana Cardoso, Bruno Martins, and Jacinto Estima. 2019. *Using Recurrent Neural Networks for Toponym Resolution in Text*.

Zhiyuan Cheng, James Caverlee, and Kyumin Lee. 2010. You are where you tweet: A content-based approach to geo-locating Twitter users. In *Proceedings of the 19th ACM International Conference Information and Knowledge Management (CIKM 2010)*, Toronto, Canada.

Grant DeLozier, Jason Baldridge, and Loretta London. 2015. Gazetteer-independent toponym resolution using geographic word profiles. In *Proceedings of the Twenty-Ninth AAAI Conference on Artificial Intelligence (AAAI 2015)*, pages 2382–2388, Austin, Texas. AAAI Press.

Grant DeLozier, Ben Wing, Jason Baldridge, and Scott Nesbit. 2016. Creating a novel geolocation corpus from historical texts. In *Proceedings of the 10th Linguistic Annotation Workshop held in conjunction with ACL 2016 (LAW-X 2016)*, pages 188–198, Berlin, Germany. Association for Computational Linguistics.

Jacob Devlin, Ming-Wei Chang, Kenton Lee, and Kristina Toutanova. 2018. BERT: pre-training of deep bidirectional transformers for language understanding. *CoRR*, abs/1810.04805.

Duarte Dias, Ivo Anastácio, and Bruno Martins. 2012. Geocodificação de documentos textuais com classificadores hierárquicos baseados em modelos de linguagem. *Linguamática*, 4(2):13–25.

Jacob Eisenstein, Brendan O'Connor, Noah A. Smith, and Eric P. Xing. 2010. A latent variable model for geographic lexical variation. In *Proceedings of the 2010 Conference on Empirical Methods in Natural Language Processing*, pages 1277–1287, Cambridge, MA. Association for Computational Linguistics.

Tommaso Fornaciari and Dirk Hovy. 2019a. Geolocation with attention-based multitask learning models. In *Proceedings of the 5th Workshop on Noisy User-generated Text (W-NUT 2019)*, pages 217–223, Hong Kong, China. Association for Computational Linguistics.

Tommaso Fornaciari and Dirk Hovy. 2019b. Identifying linguistic areas for geolocation. In *Proceedings of the 5th Workshop on Noisy User-generated Text (W-NUT 2019)*, pages 231–236.

Milan Gritta, Mohammad Taher Pilehvar, and Nigel Collier. 2018a. Which melbourne? augmenting geocoding with maps. In *Proceedings of the 56th Annual Meeting of the Association for Computational Linguistics (ACL 2018)*, pages 1285–1296, Stroudsburg, Pennsylvania.

Milan Gritta, Mohammad Taher Pilehvar, Nut Limsopatham, and Nigel Collier. 2018b. What's missing in geographical parsing? *Language Resources and Evaluation*, 52(2):603–623.

Claire Grover, Richard P. Tobin, Kate Byrne, Matthew Woollard, James Reid, Stuart Dunn, and Julian Ball. 2010. Use of the edinburgh geoparser for georeferencing digitized historical collections. *Philosophical Transactions of the Royal Society A: Mathematical, Physical and Engineering Sciences*, 368:3875–3889.

Bo Han, Paul Cook, and Timothy Baldwin. 2014. Text-based Twitter user geolocation prediction. *Journal of Artificial Intelligence Research*, 49(1):451–500.

Mans Hulden, Miikka Silfverberg, and Jerid Francom. 2015. Kernel density estimation for text-based geolocation. In *Proceedings of the Twenty-Ninth AAAI Conference on Artificial Intelligence (AAAI 2015)*, pages 145–150, Austin, Texas.

David Jurgens, Tyler Finethy, James McCorriston, Yi Tian Xu, and Derek Ruths. 2015. Geolocation prediction in twitter using social networks: A critical analysis and review of current practice. In *ICWSM*.

Ehsan Kamalloo and Davood Rafiei. 2018. A coherent unsupervised model for toponym resolution. pages 1287–1296.

Morteza Karimzadeh, Wenyi Huang, Siddhartha Banerjee, Jan Oliver Wallgrün, Frank Hardisty, Scott Pezanowski, Prasenjit Mitra, and Alan M. MacEachren. 2013. GeoTxt: a web API to leverage place references in text. In *Proceedings of the 7th Workshop on Geographic Information Retrieval (GIR 2013)*.

Diederik P. Kingma and Jimmy Ba. 2015. Adam: A method for stochastic optimization. In *Proceedings of the 3rd International Conference on Learning Representations (ICLR 2015)*, San Diego, California.

Olivier Van Laere, Steven Schockaert, Vlad Tanasescu, Bart Dhoedt, and Christopher Jones. 2014. Georeferencing wikipedia documents using data from social media sources. *ACM Transactions on Information Systems*, pages 1–32.

Jochen L. Leidner. 2007. Toponym resolution in text: annotation, evaluation and applications of spatial grounding. *SIGIR Forum*, 41:124–126.

M.D. Lieberman, Hanan Samet, and Jagan Sankaranarayanan. 2010. Geotagging with local lexicons to build indexes for textually-specified spatial data. In *2010 IEEE 26th International Conference on Data Engineering (ICDE 2010)*, pages 201 – 212.

Bruno Martins, Francisco J. López-Pellicer, and Dirk Ahlers. 2015. Expanding the utility of geospatial knowledge bases by linking concepts to wikitext and to polygonal boundaries. In *GIR '15*.

Fernando Melo and Bruno Martins. 2015. Geocoding textual documents through the usage of hierarchical classifiers. In *Proceedings of the 9th Workshop on Geographic Information Retrieval (GIR 15)*, Paris, France.

Fernando Melo and Bruno Martins. 2017. Automated geocoding of textual documents: A survey of current approaches. *Transactions in GIS*, 21(1):3–38.

Frederic Morin and Yoshua Bengio. 2005. Hierarchical probabilistic neural network language model. In *Proceedings of the 10th International Workshop on Artificial Intelligence and Statistics (AISTATS 2005)*, pages 246–252.

Jeffrey Pennington, Richard Socher, and Christopher D. Manning. 2014. Glove: Global vectors for word representation. In *Proceedings of the 2014 Conference on Empirical Methods in Natural Language Processing (EMNLP 2014)*, pages 1532–1543, Doha, Qatar.

Reed Priedhorsky, Aron Culotta, and Sara Y. Del Valle. 2014. Inferring the origin locations of tweets with quantitative confidence. pages 1523–1536. Conference on Computer-Supported Cooperative Work.

Afshin Rahimi, Timothy Baldwin, and Trevor Cohn. 2017. Continuous representation of location for geolocation and lexical dialectology using mixture density networks. In *Proceedings of the 2017 Conference on Empirical Methods in Natural Language Processing*, pages 167–176, Copenhagen, Denmark. Association for Computational Linguistics.

Afshin Rahimi, Trevor Cohn, and Timothy Baldwin. 2016. pigeo: A Python geotagging tool. In *Proceedings of ACL-2016 System Demonstrations*, pages 127–132, Berlin, Germany.

Afshin Rahimi, Duy Vu, Trevor Cohn, and Timothy Baldwin. 2015. Exploiting text and network context for geolocation of social media users. In *Proceedings of the 2015 Conference of the North American Chapter of the Association for Computational Linguistics: Human Language Technologies (NAACL HLT 2015)*, pages 1362–1367, Denver, Colorado.

Stephen Roller, Michael Speriosu, Sarat Rallapalli, Benjamin Wing, and Jason Baldridge. 2012. Supervised text-based geolocation using language models on an adaptive grid. In *Proceedings of the 2012 Joint Conference on Empirical Methods in Natural Language Processing and Computational Natural Language Learning (EMNLP-CONLL 2012)*, page 1500–1510, Jeju, Korea.

João Santos, Ivo Anastácio, and Bruno Martins. 2015. Using machine learning methods for disambiguating place references in textual documents. *GeoJournal*, 80(3):375–392.

Pavel Serdyukov, Vanessa Murdock, and Roelof van Zwol. 2009. Placing flickr photos on a map. In *Proceedings of the 32nd International ACM SIGIR Conference on Research and Development in Information Retrieval*, New York, NY, USA. Association for Computing Machinery.

David A. Smith and Gregory Crane. 2001. Disambiguating geographic names in a historical digital library. In *Proceedings of the 5th European Conference on Research and Advanced Technology for Digital Libraries (ECDL 2001)*, pages 127–136, Berlin, Heidelberg.

Michael Speriosu and Jason Baldridge. 2013. Text-driven toponym resolution using indirect supervision. In *Proceedings of the 51st Annual Meeting of the Association for Computational Linguistics (ACL 2013)*, pages 1466–1476, Sofia, Bulgaria.

Alexander Szalay, Jim Gray, George Fekete, Peter Kunszt, Peter Kukol, and Ani Thakar. 2007. Indexing the sphere with the hierarchical triangular mesh.

Richard Tobin, Claire Grover, Kate Byrne, James Reid, and Jo Walsh. 2010. Evaluation of georeferencing. In *Proceedings of the 6th Workshop on Geographic Information Retrieval (GIR 2010)*, Zurich, Switzerland.

Benjamin Wing and Jason Baldridge. 2014. Hierarchical discriminative classification for text-based geolocation. In *Proceedings of the 2014 Conference on Empirical Methods in Natural Language Processing (EMNLP 2014)*, pages 336–348, Dohar, Qatar.

Benjamin P. Wing and Jason Baldridge. 2011. Simple supervised document geolocation with geodesic grids. In *Proceedings of the 49th Annual Meeting of the Association for Computational Linguistics: Human Language Technologies (ACL-HLT 2011)*, pages 955–964, Portland, Oregon.

Wei Zhang and Judith Gelernter. 2014. Geocoding location expressions in Twitter messages: A preference learning method. *Journal of Spatial Information Science*, 9.

Author Index

Association for Computational Linguistics
209 N. Eighth Street
Stroudsburg, Pennsylvania 18360

ISBN 978-1-7138-3386-4

9 781713 833864

CURRAN ASSOCIATES INC.
proceedings.com